INSTRUCTOR 1

INSTRUCTOR 1

FOR FIRE AND EMERGENCY SERVICES

Phil Jose

Fire Engineering®
BOOKS & VIDEOS

Disclaimer

The recommendations, advice, descriptions, and methods in this book are presented solely for educational purposes. The author and publisher assume no liability whatsoever for any loss or damage that results from the use of any of the material in this book. Use of the material in this book is solely at the risk of the user.

Copyright © 2023 by
Fire Engineering Books & Videos
110 S. Hartford Ave., Suite 200
Tulsa, Oklahoma 74120 USA

800.752.9764
+1.918.831.9421
info@fireengineeringbooks.com
www.FireEngineeringBooks.com

Senior Vice President: Eric Schlett
Vice President, Educational Director: Bobby Halton
Vice President: Amanda Champion
Executive Editor: Diane Rothschild
Operations Manager: Holly Fournier
Sales Manager: Joshua Neal
Managing Editor: Mark Haugh
Production Manager: Tony Quinn
Developmental Editor: Chris Barton
Cover Designer: Jared Hood
Book Designer: Robert Kern, TIPS Publishing Services, Carrboro, NC

Library of Congress Cataloging-in-Publication Data Available on Request

ISBN 978-1-559370-578-7

Printed in the United States of America

1 2 3 4 5   26 25 24 23

DEDICATION

Thank you to my wife who made this book both better and possible. Thank you to my kids Maria and Ian. The lives you are living are an inspiration. Thank you Michael E. Thanks also to Steve, Mike, Casey, and every member, past and present, of the Seattle Fire Department.

For Aiden

CONTENTS

NFPA 1041 JPRs and Chapter Correlations ix

Preface xiii

Introduction xix

1 Elements of a Lesson Plan 1
    Where Are the Lesson Plans? 1
    Components of a Lesson Plan 2
    The Four-Step Instructional Method 8
    Summary 20

2 Laws and Principles of Learning 23
    Learning Styles 23
    Factors That Influence Learning 27
    Summary 41
    References 41

3 Lecture and Illustrated Lecture 43
    The Four-Step Communications Process 44
    Beyond the Four-Step Method 46
    Impact of Cultural Differences 48
    The Alphabet of Generations 54

4 Classroom Positioning and Setups, Audiovisual
    Equipment, and Transitions Between Media 57
    Classroom Positioning 57
    Classroom Setups 60
    Audiovisual Equipment 64
    Learning Environments 70
    Transitioning Between Media 73
    Summary 74

5   Adapting Lesson Plans, Individualized Instruction, Mentoring, and Coaching 75

Adapting Lesson Plans 75

Individualized Instruction 76

Mentoring and Coaching..................................... 78

Summary 80

6   Computer-Based and Distance Learning 81

Online Learning... 82

The Instructor's Role 86

Summary 90

7   Evaluation and Testing 91

Testing Policies and Procedures 92

Testing Bias 92

Types of Testing.. 92

Grading and Submitting Paperwork........................ 101

Summary 102

Reference 102

Index 103

NFPA 1041 JPRS AND CHAPTER CORRELATIONS

JPR Number	Chapters	Topic
4.2.2	1	Assemble course materials, given a specific topic, so that the lesson plan and all materials, resources, and equipment needed to deliver the lesson are obtained.
4.2.3	1	Prepare requests for resources, given training goals and current resources, so that the resources required to meet training goals are identified and documented.
4.2.4	1	Schedule single instructional sessions, given a training assignment, AHJ scheduling procedures, instructional resources, facilities, and timeline for delivery, so that the specified sessions are delivered according to AHJ procedure.
4.2.5	1, 7	Complete training records and reports, given policies and procedures, so that required reports are accurate and submitted in accordance with the procedures.
4.3.2	1, 2, 3, 5	Review instructional materials, given the materials for a specific topic, target audience, learner characteristics, and learning environment, so that elements of the lesson plan, learning environment, and resources that need adaptation are identified.

JPR Number	Chapters	Topic
4.3.3	5	Adapt a prepared lesson plan, given course materials and an assignment, so that the needs of the student and the objectives of the lesson plan are achieved.
4.4.2	3, 4, 5	Organize the learning environment, given a facility and an assignment, so that lighting, distractions, climate control or weather, noise control, seating, audiovisual equipment, teaching aids, and safety are addressed.
4.4.3	2, 3, 4, 6, 7	Present and adjust prepared lessons, given a prepared lesson plan that specifies the presentation method(s), so that the method(s) indicated in the plan are used and the stated objectives or learning outcomes are achieved, applicable safety standards and practices are followed, and risks are addressed.
4.4.4	2, 4, 5	Adjust to differences in learner characteristics, abilities, cultures, and behaviors, given the instructional environment, so that lesson objectives are accomplished, disruptive behavior is addressed, and a safe and positive learning environment is maintained.
4.4.5	3, 4	Operate instructional technology tools and demonstration devices, given a learning environment and equipment, so that the equipment functions, the intended objectives are presented, and transitions between media and other parts of the presentation are accomplished.
4.5.2	7	Administer oral, written, and performance tests, given the lesson plan, evaluation instruments, and evaluation procedures of the AHJ, so that bias or discrimination is eliminated, the testing is conducted according to procedures, and the security of the materials is maintained.

JPR Number	Chapters	Topic
4.5.3	7	Grade student oral, written, or performance tests, given class answer sheets or skills checklists and appropriate answer keys, so the examinations are accurately graded and properly secured.
4.5.4	7	Report test results, given a set of test answer sheets or skills checklists, a report form, and policies and procedures for reporting, so that the results are accurately recorded, the forms are forwarded according to procedure, and unusual circumstances are reported.
4.5.5	5, 7	Provide evaluation feedback to students, given evaluation data, so that the feedback is timely; specific enough for the student to make efforts to modify behavior; and objective, clear, and relevant; also include suggestions based on the data.

PREFACE

UNDERSTANDING THE NATIONAL FIRE PROTECTION ASSOCIATION STANDARDS AND NFPA 1041: STANDARD FOR FIRE SERVICE INSTRUCTOR PROFESSIONAL QUALIFICATIONS

As an instructor, you should be familiar with National Fire Protection Association (NFPA) standards, how they apply to the courses you will teach, and how they define the content of this book for the Instructor 1. NFPA standards define the specific work and performance goals required for certifications in the fire service.

Each NFPA standard outlines the duties and job performance requirements (JPRs) specific to the topic covered by that standard. For example, initial entry into the fire service is often made through training to become a Firefighter 1, progressing to Firefighter 2. The standard for these qualifications is *NFPA 1001: Standard for Fire Fighter Professional Qualifications*. When you are teaching entry-level firefighters or reviewing basic skills for tenured firefighters, the learning objectives will likely be drawn from NFPA 1001.

The duties and JPRs for the fire service instructor are identified in *NFPA 1041: Standard for Fire and Emergency Service Instructor Professional Qualifications*. Like NFPA 1001 identifies the duties and JPRs necessary for Firefighter 1 and Firefighter 2, NFPA 1041 identifies the duties and JPRs for the Instructor 1 or Instructor 2. This book is designed to help you understand the knowledge, skills, and abilities you must demonstrate as a Fire Service Instructor 1, according to chapter 4 of NFPA 1041 Standard for Fire Instructor 1.

Additionally, you will see **Points of Performance** throughout, which highlight methods demonstrated to be effective when preparing for or teaching a class in the fire service. Let's get started.

You, the Duties, and the Job Performance Requirements of the Instructor 1

NFPA defines duty as "a major subdivision of the work performed by one individual" (NFPA, 2019). In this case, that individual is you, the developing fire service instructor. You are in pursuit of the Fire Service Instructor 1 certification. When reading about a duty, accept it as a fire service industry standard performance expectation. NFPA 1041 is an established set of responsibilities you have in your role of Fire Service Instructor 1. A list of all Fire Service Instructor 1 duties is included at the end of this chapter. Chapters begin with a list of the specific duties of the Instructor 1 covered within the chapter.

> Duty: "A major subdivision of the work performed by one individual" (NFPA, 2019)

NFPA defines a JPR as "A written statement that describes a specific job task, lists the items necessary to complete the task, and defines

---

Program Management

4.2 Program Management
4.2.1 Definition of Duty. The management of basic resources, records, and reports essential to the instructional process.

4.3 Instructional Development
4.3.1 Definition of Duty. The review and adaptation of prepared instructional materials.

4.4 Instructional Delivery
4.4.1 Definition of Duty. The delivery of instructional sessions utilizing prepared course materials.

4.5 Evaluation and Testing
4.5.1 Definition of Duty. The administration and grading of student evaluation instruments.

measurable or observable outcomes and evaluation areas for the specific task." (NFPA, 2019).

Since you are on the path to becoming a Fire Service Instructor 1, let's look at the definition from NFPA 1041. A Fire Service Instructor 1 is

> a fire service instructor who has demonstrated the knowledge and ability to deliver instruction effectively from a prepared lesson plan, including instructional aids and evaluation instruments; adapt lesson plans to the unique requirements of the students and AHJ; organize the learning environment so that learning and safety are maximized; and meet the record-keeping requirements of the AHJ. (NFPA, 2019)

Once you understand how the duties and JPRs are broken down, the rest of the book will cover each duty and JPR in detail so that you can achieve certification as an Instructor 1.

Let's divide the above definition of a JPR into five parts and examine each part to understand the purpose of this book more fully and the goals you should have in learning to be an Instructor 1.

Breaking Down the JPR for Instructor 1

Part 1

"A fire service instructor who has demonstrated the knowledge and ability to deliver instruction effectively"

Notice that this JPR includes two things you must possess as an instructor. The first is the knowledge to deliver instruction effectively.

The Instructor 1, and indeed any instructor, must have sufficient mastery of the subject that they can be an effective teacher. As a developing instructor, it is incumbent upon you to recognize your duty to prepare yourself for each classroom or training ground course you intend to teach. Recognize and accept your responsibility to possess the requisite knowledge before you step in front of a group of students.

The second item in the JPR is the ability to deliver instruction effectively. You are likely pursuing certification as an Instructor 1 because you have achieved some mastery of fire service knowledge, skills, and abilities as a practitioner. You are now transitioning from the role of a practitioner to the role of teacher. There are a lot of people who are good at firefighting. Those same people may not possess the ability to teach those skills to someone else. As you have accepted the role of the teacher, dedicate yourself to the path of learning how to teach, how students learn, and how to adapt your procedures and the curriculum for your students' success. Teaching is a rewarding experience, but it does not always come easy. The constant practice of learning to be better is as applicable to the craft of teaching as it is to the craft of firefighting.

> **Job performance requirement (JPR): "A written statement that describes a specific job task, lists the items necessary to complete the task, and defines measurable or observable outcomes and evaluation areas for the specific task"** (NFPA, 2019)

Part 2

"From a prepared lesson plan, including instructional aids and evaluation instruments"

This part of the JPR identifies that, as an Instructor 1, you will be using lesson plans, instructional aids, and evaluation instruments created by someone qualified as an Instructor 2. The lesson plans you use were likely created as part of a purchased curriculum or designed by the jurisdiction you are teaching for. Each component of the lesson plan likely correlates to the duties and JPRs of the NFPA standard that is the subject of the lesson plan. This is one of the reasons it is important you understand how the NFPA Standards, duties, and JPRs relate to the subject you will be teaching.

The evaluation instruments for each duty and JPR will directly correlate to the lesson plan and describe what your students must be capable of at the end of the session. While preparing to become an Instructor 1, familiarize yourself with the evaluation tools that will be used to measure students' performance. Understanding the tools used to evaluate each student for each class is important to your students' success. You must understand the knowledge, skills, and abilities they will be required to demonstrate at the end of the training session in order to learn how to effectively teach them what they need to know.

Part 3

"Adapt lesson plans to the unique requirements of the students and authority having jurisdiction"

With each lesson plan, each student, and each venue there will inevitably be some variance in course delivery. One skill of the Fire Service Instructor 1 is the ability to adjust the lesson plan and the delivery to the specific needs of the student, the venue, or the authority having jurisdiction (AHJ). The AHJ is likely the agency hosting your training session. Variance is often related to items such as specific tools or procedures that may be employed differently across AHJs within a region. One example is the use of multiple types of forcible entry tools across the country. The lesson plan may not exactly match the tools available within an AHJ. The Instructor 1, in cooperation with the AHJ, must have the ability to adapt the lesson plan to the tools available while still meeting the minimum requirements of the duties and JPRs.

Furthermore, you must demonstrate the ability to adapt your teaching methodology or style to meet the needs of each student. Helping students learn is your purpose. We will talk later in the book about various techniques to try when students struggle. Your duty is to find a way to help them learn and successfully demonstrate they have the knowledge, skills, and abilities that are the subject of the lesson.

Part 4

"Organize the learning environment so that learning and safety are maximized"

As a Fire Service Instructor 1, you are responsible for the learning environment. To the extent you control the classroom or training ground, recognize this part of the JPR requires that you take ownership of the environment. Ensure your students have a positive and productive experience so they have the maximum opportunity to learn. This book will explore strategies for success and identify common problems that occur in the classroom and training ground environments. Learning to identify and solve problems, hopefully before the class begins, will go a long way to giving your students the best opportunity to learn.

Maintaining a safe learning environment applies not just to the physical safety of the training ground or required personal protective equipment

(PPE). Safety also applies in the classroom environment, ensuring each student is treated with dignity and respect in all phases of the learning and evaluation processes.

Part 5

"And meet the record-keeping requirements of the authority having jurisdiction"

Quite simply, this means ensuring the documentation needs of the hosting agency are met through efforts such as taking roll, signing in, recording scores, and generally making sure the necessary paperwork is handled professionally. Your attention to detail in these matters is important to the agency that you are instructing for, your students, and you. Many jurisdictions require people to sign in at the beginning of a class. Sometimes this process provides students with continuing education requirements that maintain their certifications or ability to provide service to their citizens. The paperwork could include liability releases or acknowledgment of required PPE or other safety items. As the instructor, you are responsible to ensure that the paperwork gets done as defined by the AHJ.

> **Personal Protective Equipment: "The full complement of garments fire fighters are required to wear while on an emergency scene, including turnout coat, protective trousers, fire-fighting boots, fire-fighting gloves, a protective hood, self-contained breathing apparatus (SCBA), a personal alert safety system (PASS) device, and a helmet with eye protection" (NFPA, 2019)**

Summary

The goal and objective of this book are to provide you the opportunity to gain the knowledge, develop the skill, and demonstrate the ability to meet the requirements for certification as a NFPA 1041 Instructor 1. Becoming an instructor is a fun and rewarding way to serve other firefighters and officers. Rise and meet the challenge with a positive approach, and the craft of teaching will return your investment tenfold.

The duties and JPRs covered within each chapter are listed at the beginning of the chapter for your reference.

INTRODUCTION

Welcome to the world of the Fire Service Instructor 1. I wrote this book to help you become an instructor. If you are like me, you saw an opportunity to help your fellow firefighters learn new skills and information. That's where I was early in my career with the Seattle Fire Department. Like you, my department required me to have an Instructor 1 certification before they would let me teach any official classes. At that time, I didn't really know what NFPA standards or JPRs were other than my limited exposure during recruit training. What I did know was that I had achieved some measure of technical competence, I was interested in teaching, and this was the road to helping my fellow firefighters learn so we could provide better service to the citizens we were sworn to protect.

Teaching was in my blood when I came to the fire service. My mom was a teacher. In fact, when I was growing up, she ran a kindergarten in the basement of our house. I've literally been around teachers since the day I was born. My own teaching began in high school when I worked at the swimming pool on the local U.S. Navy base. I took a class to learn how to teach swimming, something I'd done all my life, but quickly recognized that skill in the water did not equate to skill as an instructor, so I began to focus on learning how to teach. I asked questions. I sought out mentors. I experimented with different techniques and evaluated what worked. I learned every day to be a better instructor, and I brought that same focus to developing my instructing skills in the fire department.

Now, I can welcome you to the fold of fire service instructors. This is just the first step of what I hope is, like mine, a long and rewarding journey. Teaching has given me the opportunity to write this book. It has given me the opportunity to teach within my department and around the world to fire departments and at conferences such as the Fire Department Instructors Conference (FDIC), and it has given me many opportunities to meet so many firefighters from so many places. I hope your journey provides you the fulfillment you seek as you learn to help others in a new way. The instructor's way.

The fire service has a long and exciting tradition of handing down knowledge, skills, and abilities. Each generation prepares the next through mentoring, training, and instruction. Firefighters, company officers, and chief officers throughout the fire service understand that their knowledge serves best when shared. The fire service instructor fosters the environment necessary for knowledge, skills, and abilities to be developed and effectively delivered. When you accept the role and the responsibility of a fire service instructor, you agree to dedicate yourself to learning the craft of teaching alongside the craft of firefighting. Each instructor should seek to build and sustain a learning culture in their organization. Welcome to the world of the fire service instructor. Dedicate yourself to learning the methods of teaching to help build the future of the fire service.

Reference

National Fire Protection Association. 2019. *NFPA 1041: Standard for Fire and Emergency Services Instructor Professional Qualifications.*

ELEMENTS OF A LESSON PLAN

There is a tool for every job, and the tool for you, the instructor, is the lesson plan. As a Fire Service Instructor 1, you will use lesson plans to help students learn. This chapter focuses on lesson plans and the related duties and job performance requirements (JPRs) of the Instructor 1. You must demonstrate the ability to understand and use a lesson plan to meet the qualifications identified in NFPA 1041. These are the duties and JPRs for an Instructor 1 related to lesson plans.

You must demonstrate the ability to

1. Understand and describe the elements or components of a lesson plan
2. Assemble the lesson plan and the required supporting material
3. Use a lesson plan
4. Adapt a lesson plan

    a. To meet the needs of the authority having jurisdiction (AHJ)
    b. To the location where you are teaching
    c. To meet the needs of your students

Where Are the Lesson Plans?

Each class you teach will be hosted by an AHJ. The AHJ for a new instructor is typically the fire department where the Instructor 1 volunteers or works. Sometimes new instructors begin at a local, regional, or state training academy. Regardless of where your first teaching opportunity will be, the hosting AHJ is responsible for providing the lesson plan. Lesson plans are often kept within a physical library, a computer database, or both. The AHJ may provide a physical copy of the lesson plan, which will generally be in a large three-ring binder. Another option is that the AHJ stores their lesson plans electronically. If so, you will need instructions on where and how to access them, including any access privileges necessary for the AHJ. Before accepting a teaching assignment, make sure you understand where and how the lesson plans are maintained, how to access them, and how they are designed.

> *Point of Performance: Have a direct contact within the AHJ hosting the class who can assist you on your first day.*

Components of a Lesson Plan

Fortunately for you, as an Instructor 1, someone else is responsible for developing the lesson plan. Lesson plans are a comprehensive outline of everything that will be needed and every learning objective that will be covered within the scope of the class. Since lesson plans are comprehensive documents, they are often very large documents of 25 or more pages. This section will reference a forcible entry lesson plan from the Instructor Guide designed to accompany *Fire Engineering's Handbook for Firefighter I and II*. Just this one lesson plan is 47 pages long and must be thoroughly reviewed before any class you teach.

Lesson plans are generally either purchased as part of a designed curriculum or developed by an Instructor 2 within the AHJ. Either way, the lesson plan will contain some, or all, of these components laid out in *NFPA 1041: Standard for Fire and Emergency Services Instructor Professional Qualifications*:

- ➢ Job title or topic
- ➢ Level of instruction
- ➢ Behavioral objectives
- ➢ Instructional materials needed
- ➢ References
- ➢ *Preparation* (or motivation)
- ➢ *Presentation*

   ➤ *Application*
   ➤ Lesson summary
   ➤ *Evaluation*
   ➤ Assignments

Note: The four italicized components include preparation, presentation, application, and evaluation. These combine to form what is commonly referred to as the four-step instructional method. We will explore the four-step method later in this text. For now, let's focus on understanding the other components of the lesson plan as outlined in NFPA 1041.

Job Title or Topic

The job title or topic describes the subject, skill, or task the lesson plan is designed to address. You will likely recognize the job titles or topics as they represent the knowledge, skills, and abilities within the craft of firefighting. Whatever class you are assigned to teach, the job title or topic should align with your subject matter expertise. Common lesson plan titles reference specific job duties; some examples include forcible entry, nozzles and hose, search, building construction, or fire behavior. Each broad category of job title will likely have a series of lesson plans designed to work together as a full curriculum on the topic (fig. 1–1).

Fig. 1-1. Students training with an instructor

For example, a nozzles and hose lesson plan series may break down to four sections:

- ➤ **Nozzles**: Including each type and size of nozzle used within the AHJ
- ➤ **Hose**: Including each type and size of hose used within the AHJ
- ➤ **Stretching hose**: Typical methods for stretching hose from an apparatus to a target objective (the fire) within the AHJ
- ➤ **Picking up and loading hose**: Typical methods for collecting hose from the fireground and loading it on the apparatus within the AHJ

Within this sample series, there are likely to be both classroom and training ground components. Each lesson plan will define whether the behavioral objectives are intended to be completed as classroom only, training ground only, or a combination that includes a classroom session followed by work on the training ground.

Each series of lesson plans will likely be numbered in the order of recommended delivery. This type of systematic approach provides the Instructor 1 and the students with an organized pattern of behavioral objectives. Graduated learning helps ensure success in obtaining the knowledge, skills, and abilities to deploy hose and operate nozzles effectively. Graduated learning is used to organize teaching and behavioral objectives to ensure they are delivered in the order they are used. An example is teaching the student to carry the ladder, then move the ladder from horizontal to vertical, then extend the ladder. Graduated learning encompasses the idea of "crawl, walk, run" learning, where students put up a ladder slowly, then with moderate speed, then at fireground speed, ensuring competence and safety at each level.

> *Point of Performance: Review the lesson plans that come before and after the lesson you are teaching to understand where your lesson falls in the students' learning.*

In most fire service applications, lesson plans are stored electronically in a computer database, in a hard-copy format, or both. The file name for each lesson will likely include the job title and often include the related NFPA standard.

> *Point of Performance: Ensure you have the right lesson plan for the assigned class.*

Level of Instruction

The Instructor 1 must understand that there are several methods to define the level of instruction. Each AHJ will likely have a unique labeling method for lesson plans. Regardless of the labeling system, the level of instruction and the relevant NFPA standard will often be included in the title or topic.

Examples of level of instruction designation include the following:

- Firefighter 1 and Firefighter 2
  - *NFPA 1001: Standard for Fire Fighter Professional Qualifications*
- Apparatus Operator
  - *NFPA 1002: Standard for Fire Apparatus Driver/Operator Professional Qualifications*
- Hazardous Materials
  - *NFPA 472: Standard for Competence of Responders to Hazardous Materials/Weapons of Mass Destruction Incidents*
  - NFPA 472 is further broken down into the three general skill levels of Awareness, Operations, and Technician.

AHJs have a lot of discretion in how to name their lesson plans. Within some AHJs, the lesson plans define the level of instruction using terms such as basic, intermediate, and advanced. Lessons may be divided by major groups, like hose or ladders, then further divided by riding positions on an apparatus such as engine position 3 or ladder position 4. While these terms may be AHJ specific, you will need a copy of the lesson plan to review and prepare before teaching.

> *Point of Performance: The lesson plan should identify the level of instruction and align with the subject matter expertise of the Instructor 1.*

Behavioral Objectives

Behavioral objectives, often referred to as learning objectives, are student-focused statements defining the knowledge, skills, and abilities the student must be able to demonstrate at the end of the lesson. When an Instructor 2 is developing a lesson plan, they begin by defining the behavioral objectives that are the focus of the intended lesson. Generally, these are aligned with both the related NFPA standards and AHJ policy (fig. 1–2).

OBJECTIVES

The student shall

- List the specialized forcible entry tools used by the fire service
- Describe the importance of choosing the correct tool for the job

Fig. 1–2. Behavioral objectives

While the behavioral objectives should be within your area of expertise, it is okay if some require you to brush up on, or update, your knowledge or skills as defined in the lesson plan. Since information changes over time, don't be surprised if the newest methods or thinking within a lesson plan is different than what you learned when you were taking introductory classes. That's okay provided you prepare effectively, embrace the new material, and are ready to teach according to the lesson plan.

Point of Performance: As you deliver a class, your understanding of the objectives will be reflected in your performance as an instructor and the performance of your students.

Behavioral objectives are generally ordered to reflect how the lesson plan developers intended they be taught. Objectives within a given lesson plan often build upon each other.

Point of Performance: While preparing to teach, review the course objectives to ensure you are prepared with the knowledge, skill, and ability to help your students meet them.

Instructional Materials Needed

As an Instructor 1, you must demonstrate the ability to assemble the instructional materials, the classroom environment (fig. 1–3), and the training ground as outlined in the lesson plan. The instructional materials necessary for each class will be clearly identified in the lesson plan.

There are many types of instructional materials including the following:

➢ PowerPoint or other electronic presentations
➢ Whiteboard drawings

> ➤ Handouts or classroom notes
> ➤ Visual aids such as maps or photos
> ➤ Tools used for demonstrations
> ➤ Designated areas of the training ground for skill-based training
> ➤ A description of a building, including building height, for skills such as ladder placement.

The Instructor 1 must know where the materials are, how to access them, and understand the time necessary to do so before the class begins.

Point of Performance: Collect and organize the instructional material at least 15 minutes before the scheduled class time.

As an Instructor 1, you must demonstrate the ability to assemble the instructional materials identified in each lesson plan. For some AHJs, this may include scheduling a classroom. You may need to fill out requisition or checkout forms for projectors, ladders, or other physical assets. Each AHJ is likely to have a unique approach to scheduling both the classroom and training ground.

Point of Performance: Ensure you understand the hosting AHJ's procedures for requisitioning, using, and returning materials outlined in the lesson plan.

MATERIALS NEEDED

- Instructor Lesson Plan for Firefighter 1 Chapter 12 Forcible Entry
- Fire Engineering Handbook for Firefighter I & II, 2013
- Course Syllabus for Jurisdiction
- Student Roster
- PowerPoint Slides for Firefighter I/II Firefighter Chapter 12, Forcible Entry
- PowerPoint Projector
- Computer
- Chalk Board/White Board/Easel Pad
- Chalk or Markers
- Selection of Forcible Entry Tools

Fig. 1-3. Instructional materials

References

When an Instructor 2 develops a lesson plan, they conduct research using resources such as books, articles, and videos related to the course topic. The Instructor 2 uses these references to ensure the lesson plan is consistent with industry standards, department practice, current AHJ policies and procedures, and the latest research in the field. The Instructor 2 will list the references for the course including documents, books, magazine articles, videos, or any other source material used to construct the lesson plan. The Instructor 2 who develops the lesson plan should be thorough in providing a list of references within the lesson plan.

> *Point of Performance: Use the list of references provided as a guide for your preparation to teach a class.*

The Four-Step Instructional Method

Common to fire service education is the concept of the four-step instructional method. The four steps include Preparation/Motivation, Presentation, Application, and Evaluation. Each of these is included as a component of the lesson plan as described by NFPA 1041. Let's explore each.

Preparation or Motivation

When discussing the concepts of preparation and motivation, recognize that they can apply in two ways. One is your individual preparation to teach a class as an Instructor 1, which is covered later in the text.

The other way to think about preparation or motivation addresses how the student approaches the class. We will consider the student here and your role in understanding the student's preparation and motivation to achieve the behavioral objectives.

When considering the students' motivation to learn, it is helpful to start with why the subject matter is valuable for firefighters providing service to their community. The lesson plan should describe items that will demonstrate why students should be motivated to learn the objectives. Motivation may relate to students' previous training or experience. Motivation may relate to your experience or training. In class, you can try to draw upon the students' experience by asking them questions about the behavioral objectives. You can relate your experience by telling a short story.

*Point of Performance: Establish the "why" of the lesson
so that students are prepared and motivated to meet the
behavioral objectives.*

Teaching the Fire Service: Hands-on Instructor, First-Time, Every-Time

When you choose to become a hands-on instructor you immediately step into the role of subject matter expert for the skills you are teaching. Firefighters such as yourself are drawn to the field of hands-on instruction through a combination of dedication to excellence in the craft of firefighting and a dedication to serve others. It takes effort and training to produce firefighters both knowledgeable and innovative in the performance of these lifesaving tasks. Becoming an instructor increases your ability to serve your students and develop their skill, thereby producing a more effective outcome for their citizens. By assuming the role of hands-on instructor, you have become a mentor to other firefighters. To improve as an instructor, you must invest some time in learning the craft of teaching alongside the craft of firefighting. Let's examine some skills that will improve your ability as a hands-on instructor for your fellow firefighters.

The craft of teaching demands a professional approach to developing the skills of the instructor as you learn how to teach in the hands-on environment. You will need to understand several things to be effective. Examples include the lesson plan, effective coaching, effective demonstration, organizing student practice, time management, positive feedback, and the logistical challenges of the training environment. Your ability to master the challenges inherent in the hands-on training environment is critical your success as a fire service instructor. Your students are depending on you. Let's get started on your path to success.

Lesson 1: The Hands-on Lesson Plan
A necessary component for effective instruction in the hands-on environment is a good working knowledge of the lesson plan and any supportive material. The classroom environment is different from the hands-on environment. As you prepare to be a hands-on instructor, there will a few necessary additions to learning the lesson plan and the learning environment. The hands-on environment does not lend itself

---

(*continued*)

to the use of a projector or the multiple pages of a written lesson plan. Hands-on training also involves a smaller group of students than does classroom training. Let's examine these and other opportunities that can help you become a great fire service hands-on instructor.

A classroom instructor can readily refer to the lesson plan while teaching. The classroom instructor also has the PowerPoint slides to help spark their instructional memory. For the hands-on instructor these tools are not readily available. It's a given that you will need to study the lesson plan before your first day. A great tool for any hands-on instructor is to create a one-page outline of the hands-on training session with only major bullet points or reminders. Take the outline and increase the font size to make it easily readable with a quick glance, print a copy, and laminate it for use in the field.

The lesson plan in example 1 is from a hands-on air management training session where the students enter a limited-visibility prop space. This 5-minute briefing, prior to the hands-on drill, involves both knowledge and skills. The knowledge outlines the scenario for the students and their operational environment. There are also three skills outlined in the lesson plan including the ability to read the pressure gauge, communicate air volume within the team, and decide to turn around so they exit the hazard area before the low-air warning alarm activates. The lesson plan can provide the hands-on instructor the opportunity to demonstrate, and have the students practice, individual and team skills prior to entering the prop. Improving their performance in the hands-on prop improves their learning of the skillset and their overall enjoyment of the class. The instructor also builds credibility and trust with the students by demonstrating their ability to improve the students' skillset and performance. Having a quick-reference laminated version of the lesson plan allows the hands-on instructor to be more effective teaching the skills outlined in the plan. This is especially true the first couple of times you are teaching new hands-on content. As you gain experience, you will rely on the notes less and less. For every class, review the content prior to the training day to ensure you do not drift in your delivery.

Lesson 2: Demonstrate One—Do One
Learning in the hands-on environment involves building skills sequentially until they are mastered enough to layer on the next skill. Individual skills build upon each other to complete a single effective task on the

fireground. The instructor understands that each skill is valuable and learning the baseline skills will improve overall performance of the larger task. The instructor should follow a practice of demonstrating a single skill, having students practice that skill, then demonstrating the next skill. Build the individual skills into a pattern from beginning to end so that the entire task can be completed effectively.

Refer to example 1 again and recognize that there are three distinct skills outlined in this hands-on lesson plan. The first skill listed is to access and read the pressure gauge. This skill may seem intrinsically easy since every firefighter checks their air gauge every day when checking their self-contained breathing apparatus (SCBA). Experience demonstrates that this skill requires practice before entering the prop. It would be a mistake to presume that the regular practice of checking the SCBA when it is on the apparatus is adequate to performing the same skill in a different environment. The change is that the firefighter as a student in the hands-on environment will be checking the gauge in the hands-on prop or 'hazard area', while wearing the SCBA, and operating in near blackout conditions. This is significantly different than checking the SCBA on the well-lit apparatus floor. The lesson plan describes the objective to "Access and read the pressure gauge" followed by a step-by-step approach that includes a gloved hand, the backlight button for gauge, and proscribing 8-10 repetitions for each student. The entire process of describing and demonstrating this skill and having the students practice takes 30-45 seconds. The value for the students is building the muscle memory to perform one skill efficiently before moving on. Teaching skills one at a time through the practice of demonstrate one—do one, will prove a useful practice for the hands-on instructor.

Lesson 3: Develop Your Demonstration
Demonstrating a skill so that others may learn is different from doing the same skill focused on completing it safely and effectively. Think through, and then practice, talking and demonstrating your way through the required skills. It is helpful if you have access to instructional videos or mentors that already teach the skill you are attempting to master. Taking and watching a video of your demonstration is another great way to improve your delivery skill.

Consider the process of demonstrating the use of a Halligan and flat-head axe for forcible entry. Demonstrating will require talking through the hand placement on the tools, tool positioning relative to

(*continued*)

the doorknob and deadbolt, the direction the hook should face and why, inward versus outward opening doors, and more specifics. Where you will point and what you will say develops through the practice of talking and moving your way through a demonstration from start to finish.

Make sure the students can hear you. When speaking, project to the student that is furthest away from you and occasionally make eye contact with each student. When performing a demonstration in front of a forcible entry door, it's predictable that you will need to turn your head away from your work so that your students can hear you talk. Failing to do so may result in students that do not hear your instructions. When you are working, don't talk. When you are talking, don't work. Explain what you are going to do, and tell them where to look and what they are watching for. In our example, presume you are about to demonstrate using a baseball swing to drive the pick of the Halligan into the doorjamb to lever the door open. You would describe your actions while looking toward the students and tell them what to watch. Then you would direct your attention to the action of swinging the Halligan. Once the Halligan is buried into the doorjamb you would return your attention to the students explaining how to lever the tool to open the door. Repeat this pattern, step by step, throughout the demonstration.

Lesson 4: Sets and Reps
For the hands-on student the most important part of the class is practicing the newly acquired skills. Controlling the allocated training time effectively, you will need to ensure that each student gets the maximum number of sets and reps possible. With each opportunity to practice you will also need time to provide coaching to increase student performance. Some students will require minimal coaching whereas other groups will need significantly more to be successful. Develop an understanding of the practice time flow for each skill you are teaching and constantly assess methods to increase opportunities for repetition and coaching.

Example: Picture yourself teaching a group of five students how to properly deploy the final 100 ft of 1¾" hose to the front door of a house. You will demonstrate how to put the hose on their shoulder, how to deploy the hose to the door, and then how to re-assemble the 100 ft section on the ground for another repetition. Once the demonstrations are over the students will have 10 minutes to practice this skill. If the hose lay to the door takes approximately 30 seconds followed by picking up and reassembling the hose, which takes about 90 seconds for each student. The total time for a single repetition is 2 minutes, giving enough time for each student to complete five repetitions. Let's consider two options for running the 10-minute practice session.

Method 1: Predict an instructor having all five students stretch their hose at the same time then pick up and reset. The instructor attempts to watch two students to provide each some coaching or encouragement based on their performance. As the group resets, the instructor will have some time to coach two of the students and each student will get their five repetitions. The instructor, if they are watching and coaching two students for each repetition, will only be able to directly watch and interact twice with each student. This practice divides the instructor's attention between two students deploying the hose simultaneously. Divided attention on the part of the instructor is likely to decrease their ability to see the details and decrease the effectiveness of coaching provided to the students.

Method 2: Using the same five-person group and the same practice session time, a better practice is to have students individually deploy their hose in a sequential fashion. Student 1 would deploy with the instructor's full attention. Coaching as the student deploys and giving a positive statement when they are finished provides excellent one-on-one attention to Student 1 during their 30-second deployment effort. As soon as Student 1 is done deploying their hose and beginning to reset, the instructor signals Student 2 to begin. The instructor moves through the students sequentially allowing Student 1 to be reset and ready for another repetition when Student 5 completes their first. The instructor continues to move through the group providing the same five repetitions for each student in the 10-minute practice session. This method provides a much better opportunity for each student to learn because of the individual nature of the instructor's attention and coaching.

Lesson 5: Instructor Positioning
Once the demonstration is complete and the practice session starts, you should have a plan that includes where you will position yourself to see what the students are doing. Your positioning should provide you the opportunity to see the student work but focus specifically on the learning points outlined in the lesson plan. You should also be close

(*continued*)

enough to intervene either physically, say by placing your hand on the shoulder of a student, or by voice so the student can hear you. For example, if you need to have a student stop for a safety reason, you would want to be in a position for them to hear your voice when you say "stop." Consider these examples.

For the forcible entry training session discussed above, the instructor should maintain the ability to see the student's hand placement on the tools and the tool placement on the door. Since this training will involve striking the Halligan with the flat-head axe you wouldn't want a student's hand on the striking surface! For this type of training, you want to be close enough tell the student to move their hand or possibly to reach out, point to the tool at the right spot, and tell them to "put your hand here."

For the hose deployment session, you would likely walk alongside the student or have them walk toward you as you watched their hand placement on the hose and nozzle combination as they deployed based on methods outlined in the lesson plan. Once they are finished, you might take a knee next to them so you can look them in the eye to provide some coaching or positive feedback.

Think about the positioning requirements of instructing a vertical ventilation operation on a pitched roof prop. Presume the students are completing the entire evolution from the placement of the ground ladders to cutting the hole in the roof. Your positioning plan includes the ability to watch them during their ground operations *and* then while operating on the roof. Your access to the roof might be to use the ladder the students place, or you could put a second ladder to the roof as part of your setup prior to the class. Your transition from the ground to the roof should put you in position to watch them as they approach the designated vertical ventilation cutting area. Example 2 is a photo of vertical ventilation training at an acquired structure. Using it as an example, you can consider the following question about the students: Do they transition off the ladder safely? Do they sound the roof as they move? Are the two members of the team coordinating their movement effectively according to the lesson plan and demonstration? Your positioning needs to offer you the ability to assess each of these steps in their learning and provide coaching when appropriate. Planning in advance where you will be during each step of the process, and how you will transition, is important to ensure you and the students are safe and successful.

These are only a few examples of the types of positioning requirements you will need to consider as a hands-on instructor. Make positioning part of your pretraining planning process. Think through your positioning during each skill focused on being in the right place to see students perform the skills, speak effectively to the students to provide coaching, and ensure that the entire process is consistent with the safety and learning criteria of the lesson plan. Be close enough for effective teaching yet out of the way for student learning.

Lesson 6: The Students' Skills

For most hands-on training courses, there will be some prerequisite skills. These are skills that the student should, repeat SHOULD, bring with them when they attend your class. Example: For a forcible entry class, using the Halligan and flat-head axe would likely list as a prerequisite skill something like: "The ability to swing a Halligan and a flat-head axe so that the target can be hit repeatedly." Why is this important to you as an instructor of a hands-on class? You cannot presume that everyone arrives with the prerequisite skills.

When thinking about how your class will proceed, it's a sound practice to allow time for an initial review of the prerequisite skills. Plus, skills degrade over time when not exercised regularly. Have the students practice basic striking skills, watching closely for the safety points and personal protective equipment (PPE) required any time striking is underway.

There are several advantages to conducting a short prerequisite skill review to improve the class and the opportunity for your students to learn. The practice session helps set the tone relative to required PPE. You can evaluate the relative skills of your students. The students will have the opportunity to practice a skill that they may not have used in a while. They also have the chance to warm up both physically and mentally to the learning challenges before them. By practicing the prerequisite skill, students can focus on building from the prerequisite skill into a learning and operating pattern that includes the new skills you are teaching. Building skills from what they know already also increases the "stickiness" of the new skills in their motor pattern.

Lesson 7: The Positive Approach

Since your goal is to teach others, it is imperative that you come with a positive approach to learning and maintain that positive attitude across

(*continued*)

the training day. In your demonstrations, and your individual coaching, it is critical to focus on the right way of doing the skills you are teaching. Instructors sometimes fall into the trap of demonstrating how things should not be done. For the student, this type of demonstration can create confusion when they are trying to put the demonstrated skills into practice. Imagine you are a student who watches the instructor demonstrate one right way and three wrong ways to perform a skill. What do you remember and what do you do? The instructor just created confusion where, likely, none existed before. As a hands-on instructor, focus on repetition of the right way to perform any skill. Better to demonstrate the right way three times than performing the skill once the right way and two times the wrong way. Do it right every time so the students can follow your example.

The positive approach extends to your coaching interaction with the students as well. Humans learn best in a positive and supportive environment. Your students will make mistakes. Your job is to correct those in a positive way to ensure they don't make the same mistake twice. Keeping your language positive will help. For example, picture yourself working alongside a student handling the Halligan bar as part of a forcible entry team. When the student puts their hands in the wrong position on the tool tell them "put your hands here" as a positive statement while you point to the right spot for their hands. This is better than saying "don't put your hand here" as a negative. When they move their hand to the right spot support them saying something like "good job, that's right where your hands go." Bringing and maintaining a positive approach improves learning and makes the class more fun.

Lesson 8: Time Management
As you develop your skillset as a hands-on instructor, you must manage the clock. The clock moves quickly while teaching any hands-on class. Novice instructors often mismanage the clock relative to the lesson plan. Your demonstrations should require a known quantity of time. Each demonstrated skill should also have a defined amount of time for each student to practice.

For our forcible entry class, this will mean students working in pairs, switching places, and getting a few attempts in each position. Presume 10 students, working in five pairs, each with three skills to practice. Consider the impact of spending one extra minute on each of the three skill sessions. If each pair should have 4 minutes to practice each skill, the total practice time is five pair of students' × three skills × 4 minutes per pair, which is 60 minutes of training. If you allow each group one extra minute for each skill, that is five pairs × three skills × 5 minutes for 75 minutes. You will be 15 minutes over or, if no extra time is available, some students will not get to practice some skills.

Understand the number of students and an approximation of how much time each pair will have to practice the skills. Managing the time, the students, and their practice for all the course objectives is a skill that must be developed. As a novice instructor, it is likely you will need to consult your watch to stay on time. As you develop expertise you will learn to manage the demonstrations, the time, and the students based on pace and flow. A sound practice is to move through all three skills a little ahead of schedule. If you get through the three skills with 10 minutes to spare you can provide the students additional set and reps of skills that they had difficulty mastering. Giving them extra practice and finishing on time is the win-win mark of a teacher pursuing craftsmanship.

Lesson 9: The Environment
The hands-on environment produces challenges you would be unlikely to experience teaching a classroom session. Apparatus moving in the vicinity, power tools, planes, trains, and automobiles are all distractions that may be present when you are teaching a hands-on class. When there is ambient noise you will have two basic options: wait, or talk over the noise. The better option when possible is to wait for the noise to pass. An apparatus driving by, or a plane overhead for example, each has a relatively short window of impact on your teaching. If the teaching site has a noise maker that is ongoing, like saws operating, you may need to talk over the noise throughout the day, which is not ideal. A sound practice is to spend a moment surveying the training ground and the training plan for other classes before the training day begins.

Lesson 10: Have Fun!
You have learned the lesson plan, developed your demonstration, and you know your positioning, timing, and flow. You are ready to bring a positive approach to teaching your first day. Your dedication to your craft and the opportunity to help others learn is appreciated. The last thing to remember is to have fun! Teaching hands-on training is an educational and rewarding experience. By dedicating yourself to learning the craft of teaching, reviewing your performance regularly, and making steady improvement, you will prove to yourself and others that you are an excellent instructor. Enjoy the journey; it's worth the effort.

Setting the tone and pace of the lesson through your ability to relate the subject matter to the students' "why" is worth the time and effort necessary at the beginning of each class. Also understand that adults generally arrive with some motivation to learn. That is why they are taking your class.

> *Point of Performance: Align the students' motivation with the behavioral objectives.*

Presentation

Presenting new information to your students is your primary purpose as an Instructor 1. Two broad categories of presentations are knowledge and skill. Let's broadly describe each category here, recognizing that both lesson plans and individual instructors often blend the two together for increased learning opportunity.

Presentation of Knowledge. The presentation of knowledge will generally be conducted in a classroom setting. The classroom setting enables the Instructor 1 to use a variety of instructional techniques in a lecture format. Knowledge can also be delivered through computer-based or distance learning.

Within the lecture format, an instructor should look and feel comfortable at the front of the room. You should consistently adjust your technique and approach to ensure students stay engaged in and learn the material. Lectures often include instructional material such as handouts or visual aids, creating a variety of stimuli for students across the lesson plan. As a Fire Service Instructor 1, you must demonstrate the ability to use a wide variety of learning aids, including visual aids, videos, maps, drawings, white board interactions, and questions to students. Each of these represent an instructional technique you will use to maintain interest and interaction with your students.

You should select, practice, and use a variety of instructional techniques. The success of each technique within a classroom will vary depending on your personality and experience combined with the course material. Each class will also have a unique group of students demanding adjustments in your approach. As you gain experience teaching a lesson plan, you will learn to use and master different techniques. You will also learn to adjust your delivery based on the students present in your class that day. Focus your effort on the students' success in mastering the behavioral objectives. Work to improve each time you step in front of a class.

Point of Performance: Use a variety of instructional techniques to keep your students engaged and help them achieve the behavioral objectives.

Presentation of Skills. The presentation of skills is commonly done through the demonstration method. Demonstrations do occur in the classroom setting but are more common on the training ground. Demonstrations require the instructor to describe a skill, perform the skill so that students can see, and then provide time for students to practice the skill demonstrated. The ability to demonstrate good technical mastery of a firefighting technique while maintaining a safe environment is an important skill to develop. Knowing and demonstrating alternative methods to achieve the learning objective, adjusting application of skills to the space or prop available, and understanding the students' ability to perform are all skills the Instructor 1 will develop.

Application

The application element of the lesson plan is when the students begin demonstrating their understanding of the subject matter. This is when you determine what the students have learned about the behavioral objectives and asses their ability to apply them. There are a broad range of instructional techniques to allow students to apply their knowledge of the behavioral objectives. These tools include quizzes, instructor questioning, student questioning, and group activities. Each is intended to provide students the opportunity to apply the knowledge or skill being taught to the work of firefighting. The application phase should be highly interactive and focused on allowing the students to demonstrate understanding. During the application phase, the Instructor 1 provides support, correction where necessary, and continual positive feedback.

Lesson Summary. The lesson summary element of the lesson plan provides a systematic method of closing out the learning process in preparation for the evaluation phase. The summary will be a review of the primary learning points as identified in the behavioral objectives. This is an opportunity for you and your students to review the goals established at the beginning of the class. The summary also provides an opportunity to reflect on or reinforce any areas of knowledge or skill where your students are unsure. Use the summary to confirm understanding and build confidence in your students as they approach the evaluation phase.

Evaluation

The evaluation element of the lesson plan provides students the opportunity to demonstrate their mastery of the knowledge or the ability to perform the skill outlined in the behavioral objectives. The evaluation phase may include written testing, watching students perform a skill, or other standards-based approaches to the evaluation of knowledge or skill. As a Fire Service Instructor 1, you must be familiar with the types of evaluation methods designed by the AHJ for the job title and level of instruction you are teaching. While we will cover evaluation tools more in chapter 7, understand that evaluation is a duty of an Instructor 1.

Assignment. The assignment element of the lesson plan provides requirements for the students to complete work on their own. These can include a range of options such as readings, videos, quizzes, or participation in online discussion groups or chat room activities. Students may be required to complete assignments before or after a classroom or demonstration session. Assignments are developed by an Instructor 2 as part of the lesson plan and designed to facilitate students' ability to achieve mastery of the behavioral objectives.

As an Instructor 1, you will need to understand the practices of the hosting AHJ regarding assignments, due dates, collection, and grading related to assignments. The Instructor 1 must follow the direction of the AHJ related to how assignments are given to and collected from students. You should also be available to help your students complete assignments before and after your scheduled courses. Your effort supports students in preparing for individual courses. Assignments also help build the knowledge, skills, and abilities of firefighters over the course of a comprehensively designed curriculum.

Summary

This chapter describes the elements or components of a lesson plan as defined in NFPA 1041. The Fire Service Instructor 1 must demonstrate an understanding of the elements of a lesson plan. While this is a comprehensive list of elements, it is important to understand that each AHJ will likely have their own lesson plan format. Most AHJs use a lesson plan format that does not include all elements. Often, elements are combined within a lesson plan. You must be able to recognize and help your students achieve the behavioral objectives identified in the lesson plans you will be using. Knowing the elements of the lesson plan must be combined with the ability to use a lesson plan, which is addressed in Chapter 2. Fire service lesson plans are often built around the four-step method of instruction, which includes preparation, presentation, application, and evaluation. In addition, knowing, adjusting, and delivering the lesson plan so that your students can demonstrate understanding of the behavioral objectives is your purpose and your duty.

LAWS AND PRINCIPLES OF LEARNING

As a Fire Service Instructor 1, you will mostly be teaching adult learners. Your students will also be firefighters. The classes you teach might be part of a Firefighter 1 curriculum. You might be doing demonstrations of hose and ladder techniques or teaching a building construction course. Regardless of the courses you teach, there is no doubt that you will, over the course of your career as an instructor, interact with a wide variety of people from different cultures, backgrounds, and motivation levels. Each of these student traits affects your teaching efforts. In this chapter, we explore learning styles as well as factors that influence learning, including motivation, cultural differences, learning disabilities, disruptive behavior such as harassment and abuse, and a host of other influences. Keep in mind that your goal in teaching any class is the same: Help your students learn. Help them meet the course objectives. Use whatever method is most effective to aid them in their pursuit of firefighting knowledge, skills, and abilities.

Learning Styles

Depending on the sources you consult, there are as many as 11 different learning styles. This text will focus on three styles of learning: visual learning, auditory learning, and kinesthetic learning. These three styles easily relate to what you see (visual), what you hear (auditory), and what you do (kinesthetic). While you are learning to be an instructor, and when you begin teaching classes, maintain an awareness that each student likely relies on a blend of learning styles. Students may not even know their own style of learning but will often demonstrate a preference for, or perform better, when the instructor uses one specific style or blend of styles suited to them. Approach each teaching opportunity with a plan to incorporate each learning style regularly during your class. Further commit to applying each learning style while teaching each learning objective. The blend will provide help maintaining good pace and flow while appealing to a broad section of learning styles in your audience.

Visual Learning

Visual learners rely on what they can see to learn, remember, catalogue, and recall information. These types of learners build mental representations of the lesson. Visual learners combine teaching aids such as maps, diagrams, or drawings from the whiteboard with information from preclass readings and lecture material. They use the visual aids to construct mental models, or representations, of the material to catalogue information. Their mental models then assist them to effectively retrieve the information later. Your use of content related photos, videos, maps, or drawings will assist the visual learner in their effort to catalogue and recall information from the class.

> Point of Performance: Constantly assess and improve your ability to effectively use visual teaching aids in a memorable way.

Auditory Learning

Auditory learners are skilled at taking what they hear and using that information to remember the course content or learning objectives. They are excellent listeners with the ability to catalogue information while listening. Auditory learners rely on the instructor to vary the tone and pace at which information is delivered. Variations in your speech pattern help auditory learners determine the value of and remember information provided.

While many fire service courses are designed to use the lecture method, as an Instructor 1, you should understand that the lecture method can be the least effective method of teaching. Students quickly disengage from the learning process when instructors do not vary the tone of voice, the pace of instruction, or otherwise liven up the style of their delivery (fig. 2–1). Use relatable stories, case histories, and other content-rich explanations to help your students connect the lecture sections of your class to the learning objectives.

Fig. 2-1. Lecturing students—even on the fireground—can be the least effective method of teaching if students disengage due to the actions of the instructor (courtesy of Tim Olk).

Kinesthetic Learning

Kinesthetic learners will achieve the course objectives best when they are directly involved in learning through doing. Kinesthetic learners in fire service courses often excel in the hands-on environment of the training ground (fig. 2–2). Conversely, they may be challenged by the standard classroom environment. Examine your teaching style and the lesson plan for opportunities to include kinesthetic learning in your delivery. Your extra effort will help kinesthetic learners meet the learning objectives.

Combining Styles and the Learning Pyramid

The learning pyramid attempts to define how much students learn based on the teaching format employed by the instructor. A quick look at the learning pyramid will probably support how you feel generally about each teaching style (fig. 2–3). Because of the simplicity of the concept and the ease with which it generally supports our own preconceived notions of teaching and learning, we might accept it at face value. Understand there is little evidence to support such a rigid interpretation of learning styles and their effectiveness. Some research exists to support the idea that the lecture method is the least effective. Teaching others is supportable as a good method to promote learning due to the preparation needed to be an effective teacher. You will find this to be true when you begin teaching.

Fig. 2-2. Students can learn by an instructor coaching them (courtesy of Tim Olk).

Fig. 2-3. The learning pyramid

There is some validity to the pyramid and its presentation, but what you should take away from the graph is that there is a sliding scale of effectiveness. You will be more effective and efficient as an instructor if you learn to vary and combine teaching styles. Your students will also be more likely to meet the behavioral objectives.

Let's say you start the class with a short session of asking students questions (discussion-based teaching, 50%), transition into content-based lecture (10%), and then facilitate a group session combining information from both for another 50%. Now, that adds up to 110% learning, which is clearly impossible. If we just took the highest number, the result would be 50% learning, which is also unlikely. The point is that by understanding and combining different methods you will achieve better retention of the learning objectives. The pyramid is a good reminder that multiple methods are likely more effective than any individual method.

> *Point of Performance: Vary your teaching style and approach regularly to help students learn more effectively.*

Factors That Influence Learning

Many factors influence your students' ability to learn in the classroom and on the training ground. This chapter will expand on some of the major factors that affect your students' ability to learn including motivation, repetition, cultural differences in learning, learning disabilities, and disruptive behaviors to include side conversations and harassment, abuse, or discrimination.

Motivation

Every student in your class will have a unique motivation for attending. This means each student arrives with a distinct "why" of learning. Discover and value the "why" that brings each student to your class and use this knowledge to adapt your delivery, helping each student achieve the learning objectives.

> *Point of Performance: Work to understand your students' motivation by taking some time before the class starts or during the opening of your class to find out why the students are there.*

Some students attend on their own initiative to improve their understanding of the craft. Some are there to meet a requirement for more opportunity within their department. These students are generally highly motivated to meet the learning objectives and enthusiastic in their approach and interaction with the instructor and other students.

There will be some students attending your classes to maintain a minimum accreditation or other mandate from their jurisdiction. While many of these students will bring a positive approach and motivation to the course like the students previously described, others will not be highly motivated. They will likely be working to meet the minimum learning objectives of the course rather than fully embracing the course with a positive attitude.

Whatever the motivation of your students, your motivation to help them meet the learning objectives of the course must remain high. Your interaction with each student will build and maintain a high level of motivation. You directly influence the students' motivation by maintaining a positive and productive learning environment.

> *Point of Performance: Your ability to make learning interesting and engaging helps students master the learning objectives.*

Repetition in Learning

Each learning objective outlined in the lesson plan will be important for the students to be able to recall later. Generally speaking, it is a good idea to cover each learning objective several times across the course of a class. Each time a learning objective is covered, you will want to change the teaching approach to facilitate multiple styles of learning blending together for your students. This blend will help students recall the information during the evaluation phase and, more importantly, over the course of their career. What you are teaching is important to their success, and you want to help them remember the information long term (fig. 2–4).

Repetition is also critical when teaching manipulative skills. There's an unattributed quote that goes, "Don't practice till you get it right, practice till you can't get it wrong."

One key to repetition on the training ground is to have the students make each repetition perfect. This will likely require their initial attempts at a new skill to be performed slowly. Each additional repetition then should increase in speed while maintaining perfect skill application. Eventually, both the skill and the speed will be there for your students. Try to avoid focusing on speed at the expense of excellence. For example, if you are teaching how to don a set of structural firefighter bunking gear, the applicable time standard per NFPA 1001 is one minute. A student should

Fig. 2–4. Firefighters conduct a ladder drill practicing basic skills to improve their fireground performance (courtesy of Tim Olk).

practice donning bunker gear very slowly, but perfectly, through a few repetitions before the time standard is even discussed. Once the student can repetitively don the gear perfectly, introduce the time standard but emphasize perfection over speed. With enough practice they will eventually be able to meet the time standard. You will be able to predict that when the student has to don the gear for real, it will be done right, which is critical for safety, and likely quickly, which helps provide better service. If you took an alternate approach focusing on speed early in the learning process, a student would likely make multiple errors in the donning process each time. Occasionally, the student would be able to complete the task both well and quickly. When it's time to don gear for a real event what will happen? Will it be quick—or right? Focus on perfect repetitions and slowly increase speed for a better learning experience, which will translate into a better actual performance when the bell hits. Another unattributed saying supports this approach: "Slow is smooth and smooth is fast."

Cultural Differences in Learning

We discussed earlier in this chapter the three learning styles: visual, auditory, and kinesthetic. Recognize that a specific learning style may be more prevalent within a specific culture. Regardless of the specific culture and learning style, the more important point to focus on is that every culture has learners that rely on each of the three learning styles. In fact, every class you teach will have a blend of learning styles among the student body.

Cultural differences can influence learning in other ways.

Group participation. A student may demonstrate less participation in group activities. Some cultures teach that it is respectful to let others speak first. As the Instructor 1, you should be attendant to who has and has not participated and give everyone an opportunity. You may have to engage these students individually during a break to get their input on the topic.

Volunteering to answer questions posed to the class. Students may have been taught as children to defer to the opinions of others. This can be especially true if there is a real or perceived power dynamic within the classroom.

Eye contact. Some cultures consider it rude to make direct eye contact with an instructor or formal leader of any kind. Some students may also be shy or introverted. It is your responsibility to look for opportunities to get these students engaged. Look for an area where you know the student is knowledgeable and direct the question to them or seek them out during a break and let them know you will be looking for them to participate in the next class section.

These are just a few examples of how student performance and inclusion may be impacted by the difference between the cultural norms present in a classroom.

Point of Performance: While cultural differences can impact the classroom environment and each student's ability to learn, your responsibility is to maintain a positive and inclusive learning environment for every student, every day.

Disabilities

Some of your students will have disabilities that can impact their learning. These can range from learning disabilities such as dyslexia (difficulties with reading), dyscalculia (difficulties with math), dysgraphia (difficulties with writing), poor vision or hearing, or other medical issues such as a history of a traumatic brain injury, an autism spectrum disorder, or mental health issues, including anxiety, depression, and posttraumatic stress disorder (PTSD).

The impact of an individual's disability on their learning varies not only by the type of disability but also by the individual.

Table 2–1 gives some general examples of specific disability characteristics.

Americans with Disabilities Act of 1990. The Americans with Disabilities Act (ADA), enacted in 1990 and amended in 2008, prohibits discrimination against individuals with disabilities with regards to employment, housing, medical care, education, and other domains. This means that for a student enrolled in a course, the institution offering that course, as well as the instructor, has a legal obligation to provide reasonable accommodations or modifications to support the learning of students with documented

Table 2–1. Disability characteristics.

Disability	Possible characteristics that could impact learning
Learning disabilities	Makes mistakes when reading aloud and repeats and pauses often Difficulty with reading comprehension Confuses math symbols and misreads numbers Poor spelling Trouble following directions Poor memory Difficulty with organization Long processing time
Traumatic brain injury	Fatigues easily Memory difficulties Emotional outbursts Slow response time Difficulty thinking clearly or concentrating Irritable or anxious
Autism spectrum disorder	Minimal or no eye contact Difficulty with figurative language Easily over-stimulated by sensory input Poor or awkward social interactions
Attention deficit/ hyperactivity disorder (ADHD)	Poor impulse control May blurt out answers or talk too much Difficulty with time management (may be late to class or with assignments) Easily distracted

disabilities. The ADA states in § 42.126.IV.12201(f) *"that reasonable modifications in policies, practices, or procedures shall be required, unless an entity can demonstrate that making such modifications in policies, practices, or procedures, including academic requirements in postsecondary education, would fundamentally alter the nature of the goods, services, facilities, privileges, advantages, or accommodations involved."*

This means the course instructor is required to make accommodations for a student with a documented disability unless it is determined that such an accommodation would change the nature or complexity of the course. For example, a reasonable accommodation for a student with memory issues and a long processing time for recall of information would be to give them additional time on a test or to complete an assignment. That accommodation does not lower expectations of the student to master the course material. It simply addresses the specific need of their particular disability. If, however, the student is unable to complete the assignment at all, even with reasonable accommodations in place and with consideration of an alternative assignment that meets the same content expectations, that would change the fundamental nature of the course requirements, and the instructor would not be obligated to provide that accommodation.

You have a responsibility to know the authority having jurisdiction's (AHJ) policies related to learning disabilities. Recognize that discrimination based on a learning disability is illegal. Both AHJs and instructors are responsible for making reasonable accommodations for people with learning disabilities.

Providing Accommodations. In general, adults with learning disabilities are likely, but not always, aware of their needs and openly communicate how you can assist them in learning. A student with poor eyesight or hearing may ask to sit at the front of the room. A student may ask for additional reading assignments or individual help meeting some learning objectives.

Possible accommodations you might be expected to provide in the classroom include the following:

- ➢ Additional time on quizzes, tests, and assignments
- ➢ Testing in a quiet room
- ➢ Providing lecture slides or notes prior to class
- ➢ Preferential seating in the classroom
- ➢ Audiobooks
- ➢ Text reader for quizzes and exams
- ➢ Verbal rather than written responses to test questions

> ➤ Alternative assignments (as long as it does not change the complexity or nature of the course content)
> ➤ Allow breaks during tests or class lectures

Remember that you have a duty to assist each student in the effort to meet the learning objectives. Most accommodations for students with learning disabilities can be easily achieved with no distraction for other students. Make every effort to meet the requests of your students within the policies of the AHJ hosting the class. If there is a conflict that will be disruptive to the class, you should take the time to deal with it respectfully before you begin.

> *Point of Performance: Have a source within the AHJ who can assist you in meeting the needs of your learning-disabled students.*

Disruptive Behavior

Students come in all shapes, sizes, motivation levels, and social skills. One thing you can be certain of as you embark on the path of the Instructor 1 is that, occasionally, a student will be disruptive. It is challenging to deal with disruptive behavior while maintaining a positive learning environment. Not all disruptions are created equal. A student who is talking to the person next to them in the class might be slightly disruptive. A student who harasses someone or engages in dangerous horseplay is much more significant. Recognizing and dealing with disruptive behavior effectively is a valuable skill.

> *Point of Performance: You must be ready and willing to deal appropriately with disruptive behavior every time you step in front of a class.*

Let's explore some examples of disruptive behavior and some recommended actions to return the classroom to a positive learning environment.

Side conversations are generally a minor annoyance for the entire student body but can be highly disruptive for students sitting near the side-talkers. If the sidetalkers are attempting to keep their voices low and engaging in sidetalk intermittently, you may be able to defer addressing the behavior directly in front of the entire class. In this case, you should speak with them briefly during a scheduled break about their behavior. Ask them

Safety Elements for Lesson Plans

by Dave Dodson[1]

While there are no specific requirements to include "key safety elements" as part of a lesson plan, history suggests that perhaps it's time to mandate it. Online research reveals that approximately 12% of all fire service injuries in the United States are training related. Further, an average of 10 line-of-duty deaths occur each year due to training activities.[2] The statistics can be telling, but what really hurts is the fact that training is a planned event. The lessons we use to map out a training activity are the playbook we use to accomplish safe incident handling. If we are committed to injury and death prevention through safe training, then it makes sense to feature that in developed lesson plans (LPs). When using previously developed or commercially available LPs that do not have safety elements included, instructors should develop and add their own as they prepare and present a lesson.

Key safety elements are just that—specific safety considerations that should be in place and communicated to help improve safety during a training activity. They should be listed prior to the four steps (or elements) of instruction—preparation, presentation, application, and evaluation—and may include specific considerations for each of the four steps.

Obviously, a sit-down classroom session will have very few, if any, key safety elements that need to be included on the LP and presented to attendees. Those LPs that include a hands-on or drill-ground activity should include written safety elements and a presentation of them (a formal safety briefing).

While the key safety elements for any given topic will be commensurate with the complexity, duration, and risks present, some overhead safety elements can apply to all hands-on training lessons:

- Required level of PPE
- Empowerment that everyone is responsible to stop unsafe acts or alter the activity for imminent injury threats
- Safety of participants has priority over skill completion
- Participant responsibility to inform the instructor of any condition or issue that could interfere with safe skills accomplishment

> ➢ Hazards associated with the specific training environment (weather, trip/fall issues, building features, tools/apparatus, traffic, etc.)

The first four bullet items above should be included on most, if not all, lessons that include hands-on activities, drills, or evaluations. Doing so may seem overly redundant, although a stronger argument can be made that the repetitive nature of safety elements helps to establish a desired safety culture for training activities. The last bullet is a flexible one that may be a short list or a complex formula of anticipated hazards based on the lesson topic.

Generally speaking, the instructor presenting a given lesson plan is responsible for making sure the class or activity is completed safely and nobody gets hurt. This is easy to say but sometimes hard to accomplish—especially for training activities that are more complex and include many participants doing multiple tasks. In these cases, a key safety element to spell out in the LP is the need to have a dedicated instructor in charge (IIC), "multiples" (assistant instructors) to help create a smaller instructor-to-participant (I/P) ratio, and even a requirement for the appointment of a safety officer (SO), separate from the IIC. New Fire Service Instructor 1s are likely to fill the "multiple" role and can gain some instructor experience under the watchful eye of the IIC and SO.

Types of lessons or training activities that can benefit from a dedicated SO, a smaller I/P ratio, or both include the following:

> ➢ Multicompany drills
> ➢ Multiagency drills
> ➢ Training sessions that include multiple skills stations that participants rotate through
> ➢ Training evolutions that require specialized skill sets or where the participants have varying levels of training (technical rescue or hazmat mix awareness, operations, and technician levels)
> ➢ In-context drills and evolutions for probationary-level trainees

Job performance requirements (JPRs) for various topics and levels of qualification may include phrases such as "as part of a team," "under

(continued)

direct supervision," or "given a team leader." These phrases are clues that the training lesson should include a key safety element that reduces the I/P ratio. While a 1:25 I/P ratio is adequate for basic knots and tool hoisting, it is woefully short for rope rescue drills (where a 1:5 ratio should be required).

In the case of live-fire training (e.g., dedicated burn building, exterior prop, acquired structure), the assignment of a separate, dedicated SO is mandatory according to *NFPA 1403: Standard on Live Fire Training Evolutions*. Similarly, some state and local laws require that a dedicated SO (separate from the IIC) be assigned for firefighter live-fire training drills. Likewise, NFPA requires an I/P ratio of 1:5 for live-fire training lessons.

If we are truly committed to reducing firefighter injury and deaths through training, then the lessons we use for the training must highlight and feature key safety elements for the given topic.

1. Dave Dodson is a fire service author, educator, and retired battalion chief. He is author of *Fire Department Incident Safety Officer* and *The Art of Reading Smoke* series.

2. The author conducted an online review of data posted by the U.S. Fire Administration and the National Fire Protection Agency. Averages were derived from a 10-year span, 2008–2017. Only hands-on, training-related deaths were included. Those deaths that occurred while traveling to and from a training activity and those that occurred during a classroom-only activity were not included. The same level of discrimination could not be applied to injuries. Accessed January 6, 2020.

to discontinue the side talk. One way to get them on board is to ask them if you can call on them to share their thoughts with the class. Follow up by addressing them early in the next session after the break.

If they are being highly disruptive, you should attempt to correct the behavior immediately. Several approaches have demonstrated success in getting a change in behavior. One method is to look directly at them and then stop talking briefly. The sound of their voice will be the only thing happening in the room, and they are likely to look up and recognize they are a distraction. Another method is to change positions in the classroom while you keep teaching. If possible, stand right next to the side talkers. Your proximity will likely cause them to discontinue the side talk. Repeat once if they continue after you reposition. If they continue, speak to them during a break (fig. 2–5).

Fig. 2–5. This instructor is positioned next to two students who are side talking.

This will draw the attention of the rest of the students their way, likely resulting in their silence. The most direct method is to stop teaching and call them out directly, saying something like, "Will you please add your thoughts for the class to consider?" This is the most confrontational method. It is critical that you attempt this type of direct approach only after using other, less obvious, methods without success. Don't worry, the other students will be happy you handled the problem, especially if you do it in a respectful way.

Harassment, Abuse, or Discrimination

If an incident of harassment, abuse, or discrimination occurs, you should stop the class and deal with it immediately. Inform the class that there will be a break. Remove or separate the parties from the class to a private area.

This type of behavior should never be allowed. These behaviors are not only unacceptable, but they are also often illegal under federal, state, and local codes. You must understand the policies of the hosting AHJ relative to this type of behavior. You also have a duty to deal with and report any instance of harassment to your AHJ contact as soon as possible.

The United States Department of Labor (Office of the Assistant Secretary for Administration & Management, n.d.) outlines the following examples of inappropriate behavior:

- ➤ discussing sexual activities;
- ➤ telling off-color jokes concerning race, sex, disability, or other protected bases;
- ➤ unnecessary touching;
- ➤ commenting on physical attributes;

What Makes a Good Hands-On Trainer?

by Aaron Fields[1]

The mark of a successful hands-on trainer is based around the success of the curriculum. Success in this case is measured by the participants' ability to understand the "what, why, when, and how" of the material.

The curriculum should be taught in such a way that the connection between the context and the technique is clearly articulated. The curriculum should be striving to construct mental algorithms that lead to recognition-primed decision-making. With this in mind, the program and material should teach a robust, rigorous application. It must be designed to create a system rather than a collection of techniques.

Hands-on instructors should have a simple system with limited variables. The technique should be coded to a specific environment, and the phrases, "figure it out when we get there," or "every fire is different," should not be used, ever. We should be instructing in a manner that builds intellectual and physical correlation and teaches standard procedures, adding form to the events—not creating chaos. Only through a plan can situational deviation be recognized and appropriate choices be made.

Too often, hands-on training is driven by an ability to learn and perform specific techniques. In fact, the curriculum should lay the foundation of the context of technique application prior to learning how one accomplishes the task. The "when" must be understood before the "how" is instructed.

Within each critical fireground skill set, the fire service should be striving to develop subject matter experts rather than the "general contractor" style of instructor. It is required for dynamic, relevant, and engaging material that the instructor be well versed in the subject matter, including the ability to demonstrate skills, correct mistakes, and point out successes. In order for this to happen, we must pick the best person for the job.

Instructors must be prepared, being able to add aspects to keep the more skilled and practiced firefighter engaged, while being able to slow down and focus on particular elements for the newer or less practiced firefighter. A well-designed training session builds upon previous sessions and sets the stage for the next session. All training must be purposeful. Sessions should be understood with scope and scale predetermined: Are we learning a new skill or practicing an old skill in isolation? Are we putting skills together into likely combinations? Or, are we allowing crews to "scrimmage," giving them free rein to deal with simulated events in totality? Often the training sessions are mixed; nevertheless, an instructor who understands the scope is more likely to be hands on or hands off with regards to instruction, at appropriate times.

Understanding time frames is critical; don't overfill the training session. Clear goals that have a priority assigned to the material give instructors a clear understanding of what has to be achieved during the session. Often instructors try to cram too much material or too many variables into their class. This is especially detrimental if the material is a variable on the same or similar tasks or techniques.

Preparation and post-session analysis are the better part of being a successful hands-on trainer. No session should begin without preparation, and a critical post-session analysis will allow instructors to continue to grow with their material and their instruction skills.

One should remember that a smile goes a long way. Approachable instructors who are willing to admit their own mistakes will be more successful. Patronization of students, regardless of experience or situation, is never acceptable. In addition, "angry" instruction is less successful, as participants become more focused on getting through the drill than learning the material.

Finally, good hands-on instructors include time for repetitions. Good instructors don't expect perfection; mistakes are part of the learning process, and good instructors expect improvement. Our industry often has a tradition of not having time for enough reps and expecting perfection right away. We must realize that skills diminish if they are not practiced, and basics become difficult if they are not drilled.

1. Aaron Fields is a firefighter for the Seattle (WA) Fire Department and teaches engine company skills at the Washington State Fire Academy and the engine company class Nozzle Forward.

> ➤ displaying sexually suggestive or racially insensitive pictures;
> ➤ using demeaning or inappropriate terms or epithets;
> ➤ using indecent gestures;
> ➤ using crude language;
> ➤ sabotaging the victim's work; and
> ➤ engaging in hostile physical conduct.

Depending on the severity, you may be able to resolve the issue by getting the offending party to recognize the transgression and apologize. A public apology may be necessary to get the class back on track. Be sure that the student who demonstrated the unacceptable behavior understands the nature and seriousness of the transgression. Also ask the student that is the subject of the behavior if an apology could be accepted. If you do not believe the offending student understands the transgression and accepts responsibility, do not facilitate an apology. Dismiss them from the class and report the behavior accordingly.

If the behavior clearly violates AHJ policy or you feel it is too egregious to allow the offending member to continue in the class, you should ask them to leave privately after you break the class. In this case, you should offer the class a short apology without opining on the transgression itself, recognize that the student has been asked to leave, and get back to teaching.

In all cases, you should follow up with the person who was the subject of the harassment, abuse, or discrimination. Work to assist them, in any way you can, to mitigate the effects of the behavior that was directed at them. You must also follow up with the reporting requirements of the AHJ. Make notes immediately after the event or, at a minimum, once you have finished teaching for the session. This type of situation is a significant risk for you and the AHJ.

Point of Performance: A phone call to your AHJ contact is a great way to get notification and documentation started.

Additional Considerations

There are also a host of other influences on student behavior to consider before and during each course. Constantly assess your students and the environment for factors that can influence the ability to learn, such as the following:

➢ Attitude: Did they come to class wanting to learn?

➢ Experience: Will your class support or challenge their experience? Do they have any experience? Can you get them to offer their experience to the group?

➢ Knowledge and education: Do they have the prerequisite knowledge for this class? Does this class build within a comprehensive curriculum? Is your language and approach understandable based on their education?

➢ Physical conditions: Is the classroom too hot? Too cold? Can they hear the instructor? Can they hear and understand videos? Are the lights too low for them to see the instructor?

➢ Competing demands for time: Some students may have to monitor radio or respond to pagers or other alerts during your class. Will these distract other students? How will you address these distractions in your class?

Summary

This chapter covered many of the laws and principles of learning as well as factors that can influence learning. As an Instructor 1, you must demonstrate the ability to understand principles such as the learning styles, motivations, and cultural differences in your classes. You have a duty to recognize and deal with disruptive behavior. You also have a duty to know and abide by the law as well as policies and procedures of the AHJ relative to learning disabilities, harassment, and abuse. You, as an Instructor 1, are in a powerful position to influence the knowledge, skills, and abilities of the fire service. This extends to the attitude and behavior you demonstrate at the front of the room. Knowing how to provide the right leadership approach to maintain a positive learning environment is critical to your students' success.

References

Americans with Disabilities Act, 42 U.S.C. § 12201 (1990). https://www.law.cornell.edu/uscode/text/42/12201.

Office of the Assistant Secretary for Administration & Management. n.d. "What Do I Need to Know About . . . Workplace Harassment." U.S. Department of Labor. https://www.dol.gov/agencies/oasam/centers-offices/civil-rights-center/internal/policies/workplace-harassment/2012.

LECTURE AND ILLUSTRATED LECTURE

The lecture format is the most common method of instruction in the classroom environment.

When you teach a class using the lecture method, create a classroom environment that is engaging for the student and achieves the learning objectives (fig. 3–1). You will have to work to keep the lecture engaging and the students involved.

Though the lecture format is the most common approach to classroom teaching, it can also be one of the least effective methods to facilitate student learning. Presenting a lecture effectively requires understanding the communications process, the impact of cultural differences in the classroom, the operation of audiovisual equipment, and the process of transitioning between different forms of media while teaching. Every message you send as an instructor should follow the four-step communications

Fig. 3–1. A large lecture class (courtesy of Tim Olk)

process and be consistent with the policies and procedures of the authority having jurisdiction (AHJ). This includes what you say and any media or handouts used in the course.

The Four-Step Communications Process

Each message has both a sender and a receiver. The communications process describes the way that a message is both sent and received. In the lecture method, the teacher is functioning primarily as the sender. The students are receivers. When the instructor is getting feedback during a class, the roles are reversed with the students becoming senders and the instructor the receiver. Regardless of the message type or direction, there is a four-step communication process that includes the encoding, transmitting, receiving, and decoding of each message (fig. 3–2).

Encoding

When teaching, you are the primary sender of messages and responsible for encoding messages. As the sender, encoding begins with the idea you want to communicate. The sender then transforms the idea into a communicable message. For the Instructor 1, this process means taking the learning objectives from the lesson plan; blending them with words, photos, or other media available as part of the lesson plan; and delivering that message to your students. The goal is to create a message that will convey the idea of the learning objective in a way that is understandable to the students.

Transmitting

Transmitting refers to the method you use to send the encoded message. In the lecture method and classroom environment, transmitting means talking directly to your students. If you are teaching a course that is broadcast on video or over the internet, the transmission is also done

Fig. 3–2. Breakout of the four-step communication process

electronically. When transmitting your message, consider volume, tone, and pace of delivery. Students in the back of the classroom need to hear the message as clearly as those in front. For a web-based class, students at home also need to be able to hear your message. For the Instructor 1, this requires you to speak loud enough that everyone can hear. Also vary your tone, pace, and volume to keep your students engaged.

Varying the tone of your voice while teaching is an important technique to keep students engaged. A very effective technique for driving home an important message is to begin with a slightly increased pace and volume, then grow quieter to make the specific point. For the students, the initial increase in volume and pace, though slight, will indicate that something important is about to happen. Your decreased volume then causes them to lean in to hear what you have to say. While you can't do this more than a couple of times during a class, you should identify some particularly important learning objectives and practice using this technique to drive home the message.

Receiving

When students hear your message, they have received the information or idea that you encoded and transmitted. Students present with the instructor can listen and see the body language and facial expressions of the instructor. Students on the internet or through audio communication may not see the instructor but can still see the PowerPoint or other visual aid while listening to the instructor. Regardless, students combine the message of what you say with the PowerPoint or other visual aids. Consider the totality of the message sent to the receiver as you prepare to teach each lecture to ensure the learning objectives are met. Understand that even though the message has been received, the student still must decode the information.

Decoding

Decoding is the process a receiver goes through to absorb, understand, and catalogue a message. You have taken an idea from the lesson plan, encoded it into a message, and transmitted it to your students, and they have received the message. The student must now decode the message. This involves changing the language and form of the message from your teaching language to the student's own vocabulary. This individualizes the message for each student to understand, accept, catalogue, and remember the content. In order to meet the course objectives, students will need to be able to repeat each message during the evaluation phase of the course.

Beyond the Four-Step Method

In addition to the four-step process, and equally critical to your success as an Instructor 1, provide opportunities to change the dynamic so your students are the sender and you are the receiver. Recognize that keeping your students engaged will require opportunities for feedback as an additional part of the communications process. Feedback gives students an opportunity to clarify their understanding of your message. Feedback also identifies your students' progress in embracing the learning objectives.

Feedback

So, you sent your message, and it was received. The students decoded the message and hopefully understood, accepted, and catalogued the information. Later, they should be able to demonstrate mastery of the learning objective contained within that message. How do you ensure that you used the four-step communication process effectively (fig. 3–3)? The answer is to get feedback from your students.

A great way to get the feedback process started is to ask your students questions about the material recently covered in the class (fig. 3–4).

> *Point of Performance: In your preparation phase, identify questions provided in the lesson plan or develop your own questions to ask while teaching.*

As you learn the material while preparing to teach, you should identify ideas and opportunities to ask your students questions while teaching. Consider where you have questions of your own that need answering and if there is material that might challenge the status quo or otherwise be controversial. Develop questions to use on a timeline to change up the pace and flow of the class, and ensure that the questions address the learning objectives. Each of these approaches will provide valuable questioning opportunities for your students.

When you ask a question, wait and listen carefully to the answer. Identify where the answer is consistent with the learning objectives or if there is a misunderstanding or incomplete understanding of the learning objective. Listen even after your student initially appears to have finished answering, as sometimes they will add additional detail if you provide the opportunity. Ask questions directed to individual students and questions that are broadcast to the group. Using questions ensures that everyone has an opportunity to participate and contribute.

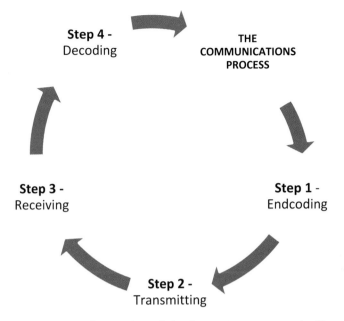

Fig. 3-3. An illustration of the four-step communication process

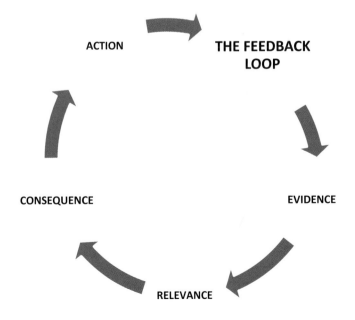

Fig. 3-4. The feedback loop

When the course is delivered over the internet, it's likely students will submit or answer questions through a chatroom or messaging platform. When a class is delivered live on the internet, students' questions are usually only accessible to the instructor. If you need to respond to questions submitted from the virtual audience, always repeat the question so students in the class and other students online understand the context of the question and the answer you provide. If you are unable to answer questions submitted online live, make sure you have a process you will use to provide answers and communicate that to your students. This is often done in writing and posted online after the class session. The important thing to remember from the feedback process is that your effort is focused on ensuring the students understand the message as it applies to the learning objectives.

Impact of Cultural Differences

The word "culture" describes the coming together of social norms, education, expectations, arts, and sciences that are accepted within a group of people. The group defining the culture can be as large as a country and as small as a family. Here, we will consider culture and its potential impact on your classroom from two points of view. One view of culture is the department culture. The other view is the country of origin or background societal culture of your students. Each of these can influence how you deliver any given lesson plan as well as how your students learn.

Department Culture

Each fire and emergency service department likely has a mission statement that includes something like, "The mission of our department is to save lives and property."

Each department also has a culture developed over decades, even centuries, influencing behavior within that department. Within the fire service more generally, the word "culture" is often used to define characteristics of large groups of firefighters, such as a "culture of extinguishment" or a "culture of search." Some departments foster a "culture of safety" or a "culture of learning" (fig. 3–5).

Individuals and departments use these labels to communicate expectations of members within the group. These can be either formal or informal expectations. Because the concept of culture is somewhat ambiguous

Fig. 3-5. A culture of safety and learning are labels communicating the expectations of its members (courtesy of Tim Olk).

in these examples, each provides both opportunity and challenge for the Instructor 1 when delivering a given lesson plan.

As you are preparing to teach a lesson plan within an AHJ, take some time during your preparation phase to consider the culture of the organization you will be teaching, how the message will be best understood, and the areas of focus or challenge the lesson plan is likely to have within that group. Your goal remains the same: Ensure your students understand and meet the learning objectives as outlined in the lesson plan.

Classroom Culture

As the instructor, you are responsible for the classroom culture you create. While each classroom might have a different group of firefighters, your approach should be consistent. You want to establish a learning culture where students have the maximum ability to learn. To do this, you must recognize some of the differences in culture outlined in this chapter. At a minimum, every classroom and training ground should be a place where students are comfortable, trusting, and safe. Mistakes are common, and they must be greeted with an enthusiasm for learning rather than ridicule or demeaning behavior. Questions should be encouraged and answered thoroughly and with respect. Establish rapport and trust with your students beginning with your first comments. Keep your language

LEARN

by Jay Dixon[1]

Teaching is an art, not a science. If it were scientific, you would know exactly how to teach any group of students any subject matter. The reality is that everyone processes information differently and learns best in distinctive ways. To complicate your job as a new fire instructor, those differences are increasing as we now find four different generations in the fire service classroom, each with varied learning styles, a challenge compounded by societal changes and technological advances.

Take the time to separate and internalize your responsibility as a classroom instructor. Whenever you find yourself struggling while teaching, come back to these core instructor values, much like we teach new firefighters to do with life safety, incident stabilization, and property conservation (LIP).

Your primary objective as an instructor is to provide your students with what they need to learn. The entire purpose of you being in the front of the room is to impart your knowledge, experience, training, and wisdom on those who sit before you. Realize, however, that the transfer of knowledge is vastly different from simply reading through and presenting the lesson of the day. You must ensure that your students understand the information and have the cognitive resources to be able to use this material as a constructive component of their new skill set.

Secondary to this objective is creating an environment that is conducive to learning for all your students. This is where the art of teaching comes in: the process of compelling your students to learn from you. While there is no defined method to achieve this, there are some constants that should be maintained any time you teach. As an instructor, you must build a rapport with your audience using the following LEARN qualities, and in turn, your students will be inclined to listen and be drawn into your teaching objectives:

Listening and open communication

Empathy

Ability and competency

Respect and trust

No lies, just honesty

Listening and Open Communication

One of the quickest ways to build trust and a healthy relationship with your students is to master the art of listening. It may be the most important active skill in teaching. It teaches you about your students and gives insight into their thoughts and questions. Listening builds trust, as it shows your students that you care about what they have to offer and you want to understand their questions. In turn, if you listen to your students, they will likely listen to you. An open line of communication built with outlined parameters creates an organized classroom where students have an understanding of boundaries yet remain comfortable and willing to communicate and ask questions. Be mindful that someone who is asking questions is engaged and curious. Maintain intellectual curiosity, as we want our firefighters to be well educated and inquisitive.

Empathy

Always look into the eyes of students while you are teaching. Think about where they are in their day. Is it a night class, and your volunteers have already worked a 10-hour day? Is it a recruit program at 07:00 after a long day of training the day before? Remember what your students may be going through, and build your environment to fulfill their needs. Teaching is not about you! Give breaks if they are tired, tell a story if they are losing interest, and change the setting to spark curiosity. Forcing information is not conducive to learning. If you were a student in your classroom, what would help you to learn better in the same situation?

Ability and Competency

If you place yourself in a position to instruct, you must be knowledgeable in that subject matter. The role of the fire service instructor is tremendous, as your lessons have the power to influence life and death situations. Ironically, students will likely remember incorrect information years down the road, as its validity will eventually collide with their continuing education. You have the responsibility to create a solid foundation for future educational building blocks.

Respect and Trust

Recognize the importance of your role as an instructor, and remember the role of those who taught you. It is vital to respect why your students are in the room and give them your best. If your students do not trust

(*continued*)

that you have their best interests, that you want to be there, or that you are providing the best information, you have lost them. They must trust that all of your intentions and information are profoundly in line with the mission of the training program. If they do not, they will likely dismiss your material.

No Lies, Just Honesty

Many people expect educators to know the answer to any question that may arise, but in reality this is not the case. No one person knows everything there is to know about a subject. If you are not honest with your answers, you will lose the trust of your audience. Remember, students can fact-check you on their phones quicker than you can finish your thought. Remain humble and be willing to do some research and follow up with the correct answer.

The above instructor traits are constant characteristics you should employ each time you teach. Using them and understanding your audience is the art. It's the hardest part of instructing, yet it can also be the most exciting.

I view the art of teaching as the process and challenge of meeting the needs of every student in the room. Remember, everyone learns differently (fig. 3–6).

Fig. 3-6. People learn differently in the classroom or on the training ground depending on how visual aids are used and where the instructor and students are located (courtesy of Tim Olk).

However, that does not mean you need to teach in a manner that is different for each student in the room. It does mean that you need to find a dynamic that crosses the boundaries for the masses. You might be wondering how to do that, and the answer may not be easily accessible. Put simply, each instructor will teach differently, and you need to find your voice and develop your own teaching persona that fosters learning.

My success as a classroom instructor is derived by employing the above instructor and environmental traits. I work tirelessly to have great communication with my students, both verbally and visually. I take body language cues to adjust my approach. For example, sitting up on the front of a chair, leaning forward, and making good eye contact implies that someone is engaged and understanding the message. Someone who looks confused is a cue for me to stop, go back, and ask what the person doesn't understand, as others may have the same question.

If students are zoned out, I walk in their direction to bring them back in or take a break if appropriate. I do not yell at or demean my students, and I always remember that I was in their shoes at one point. I limit war stories to ones that are pertinent to the conversation, and I try to create a visual picture of what I am teaching, as it may be easier for some to remember. I ask a lot of questions to measure comprehension and try to teach groupings of concepts instead of volumes of facts. Humor helps, and enthusiasm is essential.

Love your role as an instructor and your students will love coming to your classroom. Your success as an instructor will be determined by your willingness to be a great student yourself as well as your enthusiasm to continue to learn, hone your skills, and be a servant to the fire service.

1. Jay Dixon earned an education degree from St. Olaf College in 2000. He is currently a lieutenant in Torrington, Connecticut, and provides instructor development programs nationally focusing on classroom presentations and generational differences. Dixon has presented on "Death by PowerPoint: An Instructor's Guide to Educating Modern-Day Firefighters," discussing modern-day education techniques to better understand the most effective medium and instructional methods used to communicate and pass on the knowledge and traditions of the fire service. Observations show that fire service instructional styles have not advanced at the same pace as mainstream education.

professional, and make every comment positive. Stay calm, centered, and in control of yourself and the classroom or training ground. Your students deserve your respect and the respect of their fellow students. Physical and psychological safety are necessary components of a good learning environment, and that is your responsibility as the Instructor 1.

National Culture

The history of the American fire service is not the same as the vision for the future. Historically, firefighting was almost exclusively the province of White men. Even within this group, there are generational differences.

The present and the future of the fire and emergency service involves a much more inclusive approach to participation in all areas within paid, combination, and volunteer fire departments.

As an instructor in the modern fire service, you must be prepared to teach all cultures and genders. Strive to meet the needs of students as they achieve the learning objectives outlined in the lesson plan.

Different national cultures maintain different norms related to interaction with figures of authority, such as teachers. Some cultures encourage participation in a group setting while others do not. Some cultures require students to defer to the knowledge and authority of the instructor as an absolute; some allow open questioning of ideas. These norms may or may not reflect your expectations and experience as the class leader. You must be prepared to recognize, accept, and work with these cultural norms to help your students learn.

As a Fire Service Instructor 1, you must understand and follow the AHJ's policies related to cultural differences. Beyond the policies, your responsibility is to treat each student with dignity and respect, maintain an inclusive classroom, and use techniques and communication styles to provide each student the opportunity to meet the learning objectives. With each lesson plan you adopt, you have a duty to consider how you will use the communication process to effectively deliver the learning objectives to every student.

The Alphabet of Generations

Before you began your quest to become an Instructor 1, you probably heard a lot of talk about the various generations. There are a lot of labels and generalizations handed out to each generation that comes along: Gen X'ers are lazy, millennials are entitled, and baby boomers are rigid in their thinking. If you do a little digging, you can find all sorts of recommendations based on generations and quite a few that both overlap and contradict each other. While these labels might apply to some members of each group, the important thing for the Instructor 1 to consider is what impact these generalizations have on the teaching and learning process. As an Instructor 1, you should be asking, "What evidence is there that generational differences have an impact on learning?" The answer to that question is that little to no observable, measurable, and repeatable evidence demonstrates any generational impact that can be directly attributed to course content and design. Good teaching is good teaching. Remember when we covered learning styles earlier in the book and made the case that people use a blend of learning styles rather than only one at a time? Consider the same with the generational differences. Think of your students as individuals, and work hard to reach each of them directly. Help each student achieve the learning objectives through your efforts by varying the tone and pace of your lecture, using good visual aids and blending them into your lesson, and asking questions to determine understanding. These and other techniques are the backbone of great teaching. Rather than focusing on the group your student belongs to, become a great teacher for each student as an individual.

CLASSROOM POSITIONING AND SETUPS, AUDIOVISUAL EQUIPMENT, AND TRANSITIONS BETWEEN MEDIA

This chapter outlines the duties and job performance requirements (JPRs) of the Instructor 1 related to classroom setup, instructor positioning, the use of audiovisual (AV) equipment, and the transitions you make between visual aids and media. Your approach to each of these requirements has a significant impact on the learning environment and your students' success. There are a wide variety of teaching methods to address each of these topics, and this book will outline some of the most common. As you develop your skills as an Instructor 1 and as a teacher, experiment with different teaching methodologies. Pay close attention to which techniques are most effective, and build a library of skills you can rely on to help your students learn.

Classroom Positioning

Classroom positioning refers to where you stand while you are teaching a class. Understand that your position within the room will affect your students' success, engagement, and your general effectiveness as an instructor. Spend some time considering where you will stand while you are teaching the various sections of your class. Whether teaching in a classroom or on the training ground, be aware of your position relative to the students, the AV equipment, and any demonstration happening. Your students should be able to clearly see and hear you when you are teaching, so ensure your positioning provides them this opportunity. Classrooms are not all the same; your positioning must reflect how each classroom is laid out, considering everything from the light switches to the AV equipment and desk orientation (fig. 4–1).

This chapter will help you be the instructor who actively manages the classroom.

> *Point of Performance: Eye contact builds trust and improves student engagement. Establish direct eye contact with each student during the first 10 minutes of your class.*

Assess the classroom and determine three positions you will use while teaching. Each position should provide most of the students with the ability to clearly see you, avoid being in the field of any AV equipment projection, and provide you the ability to project so each student can hear (fig. 4–2).

Once you have the three positions in mind, practice teaching for 30 seconds or so from each position. Move from position to position without looking down. Take note of where the students' seats are, and predict if you will have a good line of sight to each. Now you can be confident that you can move through each position while teaching.

> *Point of Performance: Do not walk constantly while teaching. Know when you are going to move, use a relatively slow pace to move between positions, remain in the new position for at least a minute at a time, and consider moving between main instructional points.*

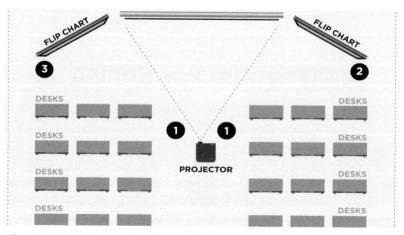

Fig. 4-1. This illustration shows three positions the instructor will be in during the class with respect to their field of vision.

Fig. 4-2. A standard classroom setup in an apparatus bay (courtesy of Tim Olk)

AV Equipment

Each authority having jurisdiction (AHJ) will have several types of AV equipment. Even within a given AHJ, different classrooms often have a variety of projectors and computers.

Arrive to the class site early enough to ensure the AV equipment is in place and works and that you have the interface tools necessary. These can include cables, adapters, wireless connections, Wi-Fi, keyboards, mice, and remote presentation pointers. Familiarize yourself with these tools before the students begin to arrive.

> *Point of Performance: Know before you start how the AV equipment works and is positioned so you avoid standing in front of the projector. If you need to move through a projection, do so quickly.*

Visual Aids

Visual aids will include photos, videos, drawings, or other items that can be projected onto the screen while teaching. Visual aids also include using a whiteboard, easel pad, or flip chart while teaching. It is important to identify what visual aids are included in the lesson plan you were provided by the AHJ. If there is a requirement that you draw on a whiteboard or easel pad, be sure to practice the drawings, at scale, in your preparation.

When using a visual aid such as a map or whiteboard, remain off to one side so that students can see (fig. 4–3).

If you have to go in front of the visual aid, to add something to a drawing for example, block the drawing for as short a time as possible. Teaching while drawing or writing at the same time is sometimes both useful and necessary, though a better habit is to do one at a time. Teach, then draw, then teach, for example. It takes practice to write or draw effectively and teach at the same time (fig. 4–4).

> *Point of Performance: Ensure you stay oriented toward the audience whenever you are speaking or raise the volume of your voice if you must speak while you are faced away. It will not help your students if they cannot hear what you are describing while using a visual aid.*

Classroom Setups

The classrooms you will use as an Instructor 1 will vary significantly over the course of your career as an instructor. You may be teaching in a

Fig. 4-3. An instructor stands beside a whiteboard while communicating with students. The instructor maintains eye contact with the students while gesturing at the board, guiding student attention to the visual aid (courtesy of Tim Olk, FDIC 2012).

Fig. 4–4. An instructor teaches in front of a whiteboard so everyone can see him speaking as well as the content on the board (courtesy of Tim Olk).

traditional classroom with individual desks in neat rows and columns. You may teach small courses at the fire station table or in an apparatus bay. Fire service instructors must have the flexibility to achieve the learning objectives of the lesson plan in a wide variety of environments. Let us discuss setup for a traditional classroom, the V and U configurations, and the online learning environment.

Traditional

In a traditional classroom configuration, each student has a desk, and the desks are arranged in rows and columns.

The same configuration may include tables instead of desks, with two to three students at each table but the tables still arranged in straight rows and columns. The instructor usually has a larger desk, lectern, or table at the front of the room, traditionally positioned to one side. In a traditional classroom, the instructor will need to alter positions regularly to maintain eye contact with the students. If you are using the three-position method described above, consider a left, right, and center approach with the left and right flanks standing within two rows of the front of the room.

> *Point of Performance: While it is OK to occasionally get close to students at the back of the room, do not spend much time teaching behind the front row. It is difficult for students to see or hear you.*

V and U Configurations

The V configuration is a variation of the traditional method. The V configuration is commonly used in a classroom with tables, though a modified version can be managed with individual desks. The V configuration will generally have a center aisle with students equally distributed on each side. Creating the V requires moving the center of the tables away from the front of the classroom and moving the ends towards the front. The desks or tables will look to be in a chevron configuration when set up correctly (fig. 4–5).

The V setup provides a better opportunity for your students to see the projector and screen, whiteboard, or other visual aids that are included in the lesson plan. The V also allows the instructor to move down the middle of the V configuration to get closer to students in the back of the room. Moving to the back of the room can be highly effective, as you are temporarily positioned close to the students there. Recognize that when you travel down the center aisle to ask a question or interact with a student, you are behind the other students. With you behind them, they will have to turn around to see or hear you and will lose the ability to see your face. You will also need to increase your volume to ensure those students behind you, at the front of the class, can hear.

The U configuration can be used with tables, individual desks, or simple chairs. It is a versatile classroom setup because the size of the U can be modified to the size of the room or the size of the audience. The setup for

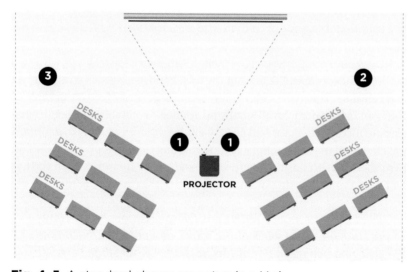

Fig. 4–5. A standard classroom setup in a V shape

the U classroom involves moving the tables or chairs into the shape of a U with the opening at the front of the room (fig. 4–6).

The U provides a less formal and more egalitarian feel to the classroom. Students can maintain good eye contact with the instructor and see each other clearly at the same time. The U configuration in a classroom session supports a highly interactive and discussion-based teaching method. The U configuration is highly effective when performing demonstrations on the training ground.

> *Point of Performance: Use the U setup for classes where you anticipate a lot of student interaction, such as a discussion-based or facilitation style of teaching.*

The Online Classroom

When teaching a class over the internet, the Instructor 1 must consider both the physical classroom environment and the virtual students. As part of your preparation, review camera locations, evaluating what areas of the classroom can be seen by a virtual student. Consider this, along with the positioning recommendations made earlier in this chapter, when determining where you will stand while teaching. When you have a combined classroom with in-person and virtual students, you must consider the impact of your positioning and classroom setup on both groups.

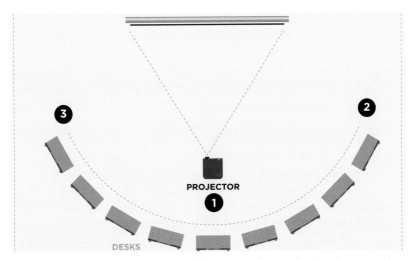

Fig. 4–6. A classroom with a table setup shown in a U shape with the instructor sitting at the head or using a stool instead of standing.

Point of Performance: Avoid positioning in areas where the virtual student cannot see you while still maintaining eye contact and interaction with the students in the physical classroom.

Audiovisual Equipment

The JPRs for the Instructor 1 include knowing how to procure, setup, use, and return AV equipment, as well as understanding the advantages and disadvantages of each type of AV equipment. You might find a classroom setup with all the AV equipment in place, or you may have to procure and return equipment. Know the expectations of the AHJ related to AV equipment in their facility. In this section, we will discuss how to use AV equipment in the classroom, including the advantages and disadvantages of AV equipment and transitions between types of media.

Advantages of AV Equipment

Using modern AV equipment enhances the classroom experience for you and your students. AV equipment also provides the lesson plan developer (an Instructor 2) a standardized format for multiple lesson plans within a fixed curriculum. The course developer has the option to blend AV learning tools. Digital slide decks such as PowerPoint or Keynote are very common. The lesson plan may also have video, photo, and audio components. This blend of learning aids allows you to present the lesson plan material consistent with each of the three learning styles. AV equipment also provides a platform to extend the classroom through the internet. It makes it possible for the AHJ to archive presentations for student review and instructor mentoring.

Point of Performance: Variable visual aids within a lesson plan allow you to alter your timing, pace, and location, increasing student engagement and enhancing learning.

Disadvantages of AV Equipment

The primary disadvantage of reliance on AV equipment is that sometimes it stops working, or you cannot get it working before the class begins. You must be familiar enough with the teaching tools you use to perform basic troubleshooting (fig. 4–7). The type of equipment will be based on the classes you teach and the AHJ you are teaching for.

Each AHJ likely has different types and qualities of AV equipment. Expect to eventually have a complete failure of the AV equipment in your classroom. While some courses require video or other visual aids to achieve the course objectives, most courses can be successfully taught by a well-prepared instructor without the use of AV equipment.

> *Point of Performance: Have a physical copy of the lesson plan with you and be so well prepared that you can use a white board and a marker if necessary.*

If a failure occurs that you cannot overcome immediately, contact the AHJ, reschedule the class, and ensure each student is notified properly.

Fig. 4-7. This visual from the head of the classroom provides an image and one word so that students will pay attention to the instructor's words and actions.

Instructor Development: Your First Classroom Session!

What makes a fire service instructor different? Can you become a fire service instructor? If so, what does it take to be a good—or even great—fire service instructor? Your first step to becoming a great fire service instructor is to make the decision to do so and then ACT!

Now that you have made the decision, great job! Congratulations, you just became a fire service instructor. The decision was the easy part; now, the action is to recognize that you have just embarked on a journey to learn the craft of teaching alongside the craft of firefighting. Recognize that even though you are an excellent firefighter, being good, or even great, at something is not a predictor of success teaching that same thing. In sports, for example, it is a well-known phenomenon that the best players do not always make the best coaches. If you want to be a great coach or a great teacher, you will need to accept that teaching is a different skill from doing. Dedicate yourself to developing the following skills to prepare for your first classroom session. This is only the beginning of the wonderful journey ahead of you as a fire service instructor! Get started today!

Lesson 1: The PowerPoint Does Not Teach! You Teach.
Your students are adults, and hopefully they can read. Teaching requires more of you than just reading the slides. Draw on your own experience as a student. You have been in that class, haven't you? You have first-hand experience in a classroom where the instructor simply read the slides. That instructor probably stood more than halfway towards the back of the room, stood facing away from the students, and likely did not maintain eye contact or interaction with the students. Think back to that experience. How effective was it? Was it enjoyable? Educational? Valuable to your skill and career development? No? Then do better. Be better. Accept as a given that the PowerPoint does not teach. You teach. Become a teacher!

Lesson 2: Be Prepared!
I know I stole that line from the Boy Scouts, but it is great advice. Collect copies of the lesson plan and the supporting documents, and begin by

reading the supporting documents. Develop an understanding of the "why" behind the lesson plan and the application. Take notes to relate the information to your own experience. Teachers know the value that an applicable story, told at the right time in the lesson, can add to understanding. Keeping track of your experience relative to the lesson plan will prove valuable.

Review the lesson plan thoroughly. Work your way through the presentation, reading the slides and the instructor notes, and regularly consult the references provided in the lesson plan. If possible, speak to the person who wrote or approved the curriculum. Watch someone teach the course while you take notes on each slide and sit down and discuss the class with them after. As you learn, take notes on the questions the lesson created in your mind, the questions you asked the other instructor, and the answers to those questions. It is likely your students will have some of those same questions. Find the answers and incorporate them into your lesson plan notes. There is always more to learn, especially if you are going to teach a topic effectively.

Example 1
Consider the lesson plan for a new standard operating guideline (SOG) within your department. You should read and reread the SOG until you can recall it almost verbatim. Do the same with any reference material cited in the SOG, which may include other SOGs, training guides, books, and articles from fire service magazines. Read them all to understand how they do and don't apply to the lesson plan you are learning.

Work to identify conflicts between the lesson plan, the supporting documents, and your understanding of the subject. Trust that if conflicts exist in your mind, your students will identify them as well. Seek out the resolution and develop your understanding so you can address questions as they arise. Having an answer is important, but, as a novice instructor, be prepared to admit you do not know the answer to a question. When this occurs, promise to get back to the student asking the question, then follow through, get an answer, and let the student and class know what it is. You build credibility by demonstrating your commitment to providing the right answer. You will lose credibility if you shoot from the hip, especially if you are wrong.

(*continued*)

Lesson 3: Practice, Lesson 4: Practice, Lesson 5: Practice

Practice your delivery. Teaching is a skill. Practice improves skill, and focused practice improves skill faster. Go through the presentation using the slides and the slide notes. Your initial practice sessions can be just a mental exercise as you think through what you will say. Do not hesitate too long before you start talking your way through the lesson. While talking, you should be standing up and practicing speaking out to your virtual students. Ask the questions in your lesson, and give space for them to answer. Talking your way through the lesson will help you build the mental models necessary for effective delivery in front of a live group of students.

Lesson 6: Timing and Flow

Working through the lesson helps you understand the timing and flow of each slide relative to the entire lesson plan. Timing and flow are best perfected through practice using the lesson plan as a guide for time spent on each section. While you are talking your way through the slides, take notes on the messaging you think will be effective. Regularly update the instructor notes version of the lesson plan while making sure you continue to meet the course objectives.

While practicing your delivery and talking through the slides, insert the questions and ideas you developed earlier in your preparation. Evaluate which will engage your audience in the material. It is likely some students already have some expertise in the lesson topic. Questions offer your students the opportunity to think about the application of the new material to their operational world. Leveraging student engagement and experience adds value to a class and helps the ideas stick in their memory. Giving them the opportunity to help reach the course objectives will also increase their buy-in to the topic.

Example 2

If we continue with the example of teaching a new SOG, it is highly likely that there are reasons why the new SOG was developed. Since you know the supporting information, you are in a great position to recognize which students in your room can add value. Identify and seek to engage those students directly with your questions. While practicing, take the opportunity to pose your questions aloud. When they sound good to you, check them against the course objectives and then include them in your lesson plan instructor notes. A good practice is to put questions in a different color in the lesson plan to highlight them during practice. Extra value is added when you develop questions that tie more than one slide together, questions that reflect the current slide topic to a previous slide, or questions that predict or forecast to slides that are soon to come.

Lesson 7: Focus on the Hard Stuff
During practice, identify sections of the lesson plan that are easier and more difficult for you while teaching. Spend extra time working through challenging sections. It may help to self-evaluate why the section is challenging. When deciding to practice a difficult section of the lesson, begin two or three slides before, and carry your practice past the problem section. Transitioning in and out of a problem section builds a good mental model representing how the section fits in the overall lesson. This type of practice also builds a reminder for you that a challenging section is coming and gives you the opportunity to focus on your delivery. This technique can also apply to sections of a lesson you find particularly valuable or that have an emotional component. By focusing some of your practice on these sections, your overall performance as the teacher will improve.

Lesson 8: The Test Teach
Now you are ready to test teach the class to a couple of "students" who can also function as mentors. You will need a few people you trust to be students who can provide good feedback after the test teach. Try to replicate the logistics of the classroom you plan to use, including the projector, lighting, and your positioning in the classroom. You should present the entire class, including breaks and handouts, to your students, and they should look at the class from a student's perspective. They should engage when you ask questions, refer to the material in handouts when you indicate they should, and generally try to be the audience you expect when you deliver the class. Once you have finished, they should provide feedback on your performance as a teacher. Their feedback will be invaluable! Use it for a few additional practice sessions before your first teaching day.

The First Day!
The first day is here! You are well prepared because you have read, developed, and practiced your delivery! Now, enjoy teaching the class. You will make mistakes. Accept this going in, and do not dwell on them.

(continued)

Just keep moving through the lesson. What is important is that you have developed your knowledge base, you understand the lesson, you practiced the delivery, you preplanned questions to ask the audience, and you can present with confidence. Your confidence will carry the class a long way even through the errors you will make.

Lesson 9: Self-Reflection

One last critical step for developing as an instructor—perform some self-reflection after every class, and occasionally have a mentor in the room watching you. Mentoring is a particularly strong improvement tool when you are starting on a new curriculum. Having someone in the room to monitor your performance and provide productive and supportive feedback immediately after the class ends is the mark of someone who wants to be better. Effective mentoring and feedback improve skill development faster. It may seem like the road to improvement is challenging, but anything worthwhile is. Being a great teacher in an area you love is priceless.

Remember, the PowerPoint does not teach! You teach. Practice and be prepared. Learn your timing and flow. Focus some energy on the hard stuff but enjoy your first day! Teaching is a skill, and skills can be developed. Coaching matters. Practice matters. The students matter. The message matters. Welcome to the world of the fire service instructor. We have been waiting for you.

Learning Environments

Every learning environment offers challenges and opportunities for the Instructor 1. Understanding and controlling the learning environment involves addressing a host of issues including lighting, distractions, climate control, and noise. Many of these are highly controllable in the classroom environment but less so on the training ground. We will discuss each issue, providing some insight to improve your teaching effectiveness and your students' opportunity to meet the learning objectives.

Lighting

In the classroom, you will likely have good control over the lighting and the ability to vary the lighting during your class. Students need to be able to see presentations on the screen as well as the whiteboard or other visual aids outlined in the lesson plan. You may need to dim or even shut off the lights when showing a video, or you may choose to decrease the lighting for effect, such as when listening to fireground audio. Review the lighting available and decide before the class begins. When you turn the lights down for effect, remember to increase the lighting before continuing the lesson. When using lighting for effect, remember to maintain eye contact.

> *Point of Performance: Remember, for most of your program you should be able to see your students and your students should be able to see you. Eye contact builds trust and improves learning.*

On the training ground, ensure there is enough lighting to provide a safe and positive learning environment. Demonstrations require that your students see what you are doing and hear what you say (fig. 4–8).

You will probably have opportunities as an Instructor 1 to teach within dark or even zero-visibility environments. This happens when you are teaching in live-fire or simulated-fire environments. These classes make teaching and learning a lot of fun, though learning to teach in this environment is a skill that takes practice. Teaching in live-fire environments usually takes additional certification beyond Instructor 1.

Never compromise your safety or the safety of the students due to lack of lighting. An instructor should have two sources of light available when

Fig. 4–8. A training ground instructor shows students, maybe taking a knee to watch (courtesy of Tim Olk).

working on the training ground. This includes building or ambient light and a flashlight maintained by the instructor.

> *Point of Performance: Keep your flashlight on but covered by your hand when operating in zero-visibility or near-zero-visibility conditions. This minimizes ambient light but allows you to immediately add light to any situation without having to find and turn on your flashlight.*

Distractions

In the modern learning environment, there are plenty of distractions, including cell phones and pagers. In the fire and emergency service environment, the Instructor 1 will also have to deal with the distraction of radio traffic for students attending while on duty as well as the house bell if teaching in a fire station.

The primary tact for the instructor is to minimize distractions as much as possible at the beginning of the class. Ask your students to turn off their cell phones and other electronic devices if possible but, at a minimum, their devices should be in silent or vibrate mode. When a phone rings during the class, simply stop teaching briefly while the student either silences the phone or leaves the room. Once the distraction is over, resume the lesson. If you are dealing with the same group over time, it may be necessary to address repeated interruptions privately rather than let a student regularly disrupt the classroom.

> *Point of Performance: Remember that everyone makes mistakes. Giving your students some grace when this happens will build trust.*

Students who are on duty may have to maintain radio contact. Students arriving with radios in hand give you an opportunity to determine what requirements they have relative to their radio and possibly responding during the class. If they are not required to maintain contact by radio, ask them to switch the radio off. If students must respond in the middle of a class, stop teaching until they are out of the room. Proceed once order is reestablished.

> *Point of Performance: Students who must maintain radio watch or may have to respond during the class should sit in the back or sides of the classroom to minimize distraction.*

Noise

In the classroom environment, noise issues can almost always be antici-pated and dealt with by proper preparation and understanding of your classroom. Noise in and around the building can normally be minimized inside the classroom by closing the doors and windows. Sometimes there is ambient noise from heating or ventilation equipment that you will not be able to address. In these cases, be prepared to speak louder or turn up your audio when the ventilation system comes on. Evaluate the volume of your presentation materials, such as video and audio clips, to ensure they are loud enough for students to hear but not overly loud. Your fellow instruc-tors will appreciate your consideration.

Point of Performance: Understand how the volume of your classroom and audio may impact instructors in adjacent rooms.

On the training ground, noise is often unavoidable. Work with your fellow instructors to understand the different courses or demonstrations happening on the training ground during your class and minimize the impact of noise for everyone operating on the training ground.

Point of Performance: When noise intrudes, such as a passing apparatus or a plane overhead, briefly stop talking until the noise passes and then continue.

Transitioning Between Media

Most lesson plans will require the Instructor 1 to transition between several types of media. Classroom presentations are often in a PowerPoint format. You may have to stop the PowerPoint to play an audio or video clip or use another visual aid such as a map or the whiteboard then transition back to the PowerPoint. Each of these transitions offer the opportunity to enhance the students' ability to meet the learning objectives. Unfortunately, each also offers an opportunity for you to lose time or credibility in your class. Practice your delivery to ensure you have a plan for each media transition.

Point of Performance: Practice going into and out of each transition during your preparation for teaching any new class.

Often video links or complete videos are embedded in the lesson plan. Embedded video and audio clips are sometimes embedded so they play with a click of the mouse. The lesson plan can also have the video links play automatically when the PowerPoint slide is advanced. The launch method should be the same for every clip within a given lesson. Know how the lesson plan is organized so that you are smooth going in and out of each transition.

It is also critical that you go through the presentation before the class to check each embedded link. Ensure that each link works, you know how it works, and the link is readily available on the PowerPoint slide. Searching for files during a class is a sure way to have students disengage.

Point of Performance: Have all the media for a given class in the same folder on the computer and have the folder open prior to the start of your class.

Summary

Classroom positioning, setup, use of AV equipment, and use of visual aids are all skills you will need to master as you develop as an Instructor 1. Student success relies on your ability to communicate effectively. Vary positioning to maintain eye contact and ensure your students can hear you. Choose the classroom setup that will best facilitate the teaching methodology you intend to use. Understand the AV equipment and visual aids for each session and practice transitioning between them so you can do so seamlessly during a class. Continue to develop each of these skills to achieve proficiency in the art of teaching.

ADAPTING LESSON PLANS, INDIVIDUALIZED INSTRUCTION, MENTORING, AND COACHING

As an Instructor 1, you will have opportunities to interact with adult learners from many cultures and backgrounds. Some students will arrive with learning disabilities or need focused, one-on-one training to be successful. These challenges and more will be yours to tackle, but each also provides an opportunity for you to hone your skills. The Instructor 1 must demonstrate the ability to adapt the lesson plan, the environment, or the teaching methodology based on student needs. Each adjustment must meet the needs of the students and ensure they can obtain the course objectives. You have the duty and the opportunity to teach, coach, and mentor students. Your tools are the lesson plan, the learning environment, and your approach to each class you teach. Let's explore some of the challenges and opportunities provided in your role as the Instructor 1.

Adapting Lesson Plans

The preparation phase for any class you teach will include a review of the lesson plan. While preparing, keep in mind the learning environment for the class. The learning environment includes the facility provided by the hosting authority having jurisdiction (AHJ). You should know and be familiar with the specific classroom or training ground area assigned. The AHJ should also provide the number and type of students attending. For example, are the students part of a recruit class? Are they tenured firefighters? Officers? Is this an introductory or maintenance class? Is the classroom completely set up, or will you have procure and set up some of the equipment when you arrive? Each of these elements helps crystallize the vision you develop for the class.

Other factors that influence your delivery include time allocated, meals or breaks during the class, and whether they occur at prescribed times or are flexible. There may be one or more students who have disruptive behaviors or learning disabilities. The number of variables that influence the success of an individual class are almost endless. The Instructor 1 should preplan and practice how to deal with variables by having action strategies in mind should something unexpected occur.

Point of Performance: Keep your mind on the twin priorities of student performance and the learning objectives.

Adapting the Lesson Plan

There are many reasons to adapt a lesson plan for a specific class or group of classes you are teaching. Table 5–1 shows examples of the types of predictable problems and projected solutions. Thinking problems through before the class allows you to develop plans and be ready for challenges that arise.

Point of Performance: When adapting a lesson plan, review your plan with a mentor.

Individualized Instruction

Fire service classes are not generally offered in a one-on-one format. There are some advantages for the student in the direct instruction environment. Consider these advantages when you have the opportunity to teach to one student directly (fig. 5–1):

> ➢ Distractions are limited in a one-on-one environment.
> ➢ The Instructor 1 can identify and use the specific learning style that the student prefers (see chapter 2).
> ➢ The time spent on course objectives can be customized to each student's preclass knowledge base.
> ➢ It provides a low-stress learning environment.
> ➢ It allows the student to volunteer answers without social pressure.
> ➢ The single student knows the instructor is listening and all needs are being addressed, leading to increased self-confidence, trust in the Instructor 1, and improvement of the learning process.

Table 5-1. Lesson plan problems and solutions.

Reason lesson plan needs to be adapted	Solution
Disruptive student	A sound strategy is to address disruption immediately and professionally to maintain a positive training environment. Disruption that includes harassment or discrimination must be addressed immediately as outlined in chapter 2. General strategies for specific disruptive behaviors are outlined in chapter 2.
Lesson plan includes group exercises but there are not enough students.	Consider an instructor-led discussion. Focus on drawing ideas from your students rather than "telling." Have students work individually to write down their ideas or impressions related to the exercise, then explore the ideas as a single group.
The breaks or lunches as outlined in the lesson plan do not coincide with the needs of the facility.	Plan breaks before the class starts. Review the lesson plan for transitions between two ideas or between media types. Find the best possible place to take breaks while meeting the course objectives and needs of the host agency. Ensure your class starts, breaks, and ends on time. Explain the plan to your students at the beginning of the day so that they know and understand the plan.
One student is not keeping up with the group.	Let the student know you will meet during a break or lunch to offer individual tutoring time. Make sure it is known that you are committed to the student's success. Follow through to help the student meet the learning objectives. For group exercises, pair the struggling student with one who clearly understands course objectives.
The time scheduled at the host agency does not match the lesson plan requirements.	Address this with your contact at the AHJ or host agency before the class date. Some agencies schedule all classes to begin and end at the same time as a matter of convenience. Understand the host agency expectations and plan accordingly to meet the course objectives.

Fig. 5–1. One-on-one instruction has many advantages.

Mentoring and Coaching

While mentoring and coaching are often used interchangeably, you should know that there are differences between them. As an Instructor 1, you should actively decide which approach you will use based on the needs of the individual student or group of students you are working with. Assessing their learning styles and the time allotted for the class may help you decide which approach will be most effective. It is likely you will often use a blend of the two techniques. Developing as an Instructor 1 requires that you actively use multiple methods to meet the course objectives while learning what works best in any given situation.

Coaching

Consider the learning process like walking down a path constructed from the learning objectives. Achieving each learning objective moves the student along the path from "not knowing" to "knowing" each concept. Coaching begins with the coach knowing the learning objectives and then leading the student down the learning path. Each student must walk a unique path to achieve the learning objectives. The student must do the work required. This includes completing the readings and exercises, as well as asking questions. The coach leads by defining the readings designed to meet the learning objectives. The coach clearly provides the direction the path is traveling, and the path leads to the learning objectives defined in the lesson plan.

Mentoring and Coaching: Being a Positive Influence

by Frank Viscuso[1]

Mentoring and coaching are two of my favorite words. I think the reason for this is because at my core, I have always tried to be a positive influence in the lives of others. You may not realize this, but you are always influencing those around you. Your words, gestures, body language, and attitude are constantly sending signals to people. The fact is that any time someone's attention is on you, you are influencing them, and as an instructor, the attention will always be on you. With that in mind, you can choose to do one of two things. You can simply instruct, or you can mentor or coach those individuals.

I have been an officer in the fire service and a certified instructor for more than 16 years. In addition to that, I've also led multiple sales teams, traveled throughout the world providing leadership and team development training to organizations, coached more than 20 Little League Baseball teams, and I'm the father of three boys. Through the years, I have had the privilege of coaching and mentoring thousands of individuals and teams. I share that with you for a reason. As a result of my experiences, I have learned that the great majority of people out there are looking for guidance and direction. They want to achieve better results in one or more areas of their life, and that is why they somehow ended up in my class or training session. That same thing can be said for every individual you will find in yours. I want to encourage you to do your best to connect with those individuals. Let them know you truly care about them and want to help them achieve their personal goals. Start coaching them. You coach by providing instruction and encouraging others to continue to move in the right direction. You correct them when correction is needed and celebrate their successes along the way.

Mentoring is a bit different. I have come to realize that we can coach anyone who is looking for a better result, but we can only mentor those who want to be mentored by us. In other words, every person will choose mentors at one point or another in their life. Think about it—you likely have people who you look up to. Perhaps the person is a coworker or someone from another organization who you hope to be like one day. Maybe you speak to that person, or maybe it is someone you hardly ever

(continued)

see, and that person has no idea. Either way, you still want to emulate some of the successful traits, characteristics, and results that person has. If that's the case, that individual may be a mentor of yours, and perhaps you just haven't used that word to describe them before now. What that person is to you, you can become to others. And I encourage you to set out to be that. Strive to be more than an instructor. Strive to be coach to some, a mentor to others, and a positive influence to everyone you come in contact with.

1. Frank Viscuso is a retired deputy chief and best-selling author of *Step Up and Lead* and *Step Up Your Teamwork*.

Mentoring

Mentoring can be thought of as walking alongside someone on a learning path. As the mentor, you take the journey with your mentee as they work to both identify and meet goals and objectives. As the mentor, you then help them discover their own path to achieving those goals. You will check in with them and provide direction and challenge them along the way. The goals and objectives may change, but you will help maintain focus and energy toward meeting the requirements of the course objectives.

Summary

The Instructor 1 must adapt lesson plans in order to meet the needs of the host agency or AHJ while also ensuring students have the opportunity to demonstrate mastery of the course objectives. There will be times when providing individualized instruction, coaching, and mentoring will assist students in developing understanding, and the Instructor 1 will need to develop each of these skills and know when they should be used. Maintaining a positive learning environment, identifying the learning style of the students and the group, and implementing the right approach to ensure student success is part of the craft of teaching and a skill set the Instructor 1 needs to be effective.

COMPUTER-BASED AND DISTANCE LEARNING

Computer-based and distance learning involve courses where the teacher and students are not located together. This type of learning uses the internet to build virtual classrooms where students access the course remotely and the instructor functions as the host. Online and distance learning help create more learning opportunities.

Though many people think of computer-based and distance learning as a recent phenomenon, it has been around for decades. Colleges and universities began experimenting with distance learning in the 1980s. The internet then was certainly not the high-speed environment we see today. With modern high-speed internet access, most universities, many high schools, and even some middle and elementary schools offer a wide variety of web-based educational opportunities to a growing number of students. The COVID-19 pandemic significantly increased both the scope and scale of computer-based and distance learning. Fire service organizations are quickly adapting to computer-based and distance learning. Departments continue to experiment with the ability for distance learning to prove an effective training tool. Distance learning allows departments to keep units in service and in their response district, and it often provides flexibility for the company officer to schedule the training within the ever-increasing demands for public service. As a new Instructor 1, you must be familiar with and prepared to teach web-based courses.

Web-based instruction is referred to by a variety of titles including online learning, blended electronic learning, web-based instruction, computer-based training, interactive television, and podcast presentations. Regardless of the title, the Instructor 1 must understand how computer-based and distance learning work. Each opportunity to teach, whether in the classroom, on the training ground, or using a web-based approach, requires that you be prepared to adapt your teaching style and the lesson plan to the environment. For computer-based or distance learning classes, you must be in close cooperation with, and getting good support from, the authority having jurisdiction (AHJ) hosting the class. Many of your Instructor 1 responsibilities such as attendance reports and testing will rely heavily, if not completely, on the systems provided by the AHJ. Familiarity with the system, navigating within it, projecting learning aids, and other skills will be critical to your success as an Instructor 1 in a distance learning environment.

Let's discuss a range of computer-based and distance learning environments while outlining the role for you, the Instructor 1, in each. Recognize that these educational models are innovating and changing rapidly. This chapter will describe some widely used methodologies with generalized descriptions.

> *Point of Performance: As with any teaching endeavor, the Instructor 1 must understand their role within the curriculum and the learning environment created by the AHJ.*

Online Learning

For this discussion, we will define online learning as a classroom environment where all the interaction takes place electronically. This includes interaction between the student and the instructor and any interaction among the students. There are two basic constructs used to organize online learning environments. These are self-paced and institution-paced courses.

Self-Paced Courses

A self-paced course allows the student to drive the course schedule based on the ability to produce required assignments. Though the class is self-paced, the AHJ generally has a required start and end date for each course. A different model allows students to choose when they have the time to complete the course. The end date provides some structure to ensure the time from beginning to completing a course is manageable for the student and the AHJ. The time frames for completing a self-paced course can be as short as an hour for some types of refresher training or as long as two to three months for collegiate-level classes.

In a self-paced course, each student completes and submits required work by the closing date. There are generally few, if any, intermediate due dates or work submission requirements. Assignments are generally submitted in the order outlined in the course syllabus with each having clearly defined content and grading standards. In self-paced courses, students choose when and how they complete assignments.

The AHJ hosting the class should clearly define expectations for both the students taking the class and the instructor to provide support and guidance. You must understand your duties as defined by the AHJ. Those duties will likely include grading and returning papers within set time frames, virtual office hours to answer emails, discussion board posts, or video chats. Keeping track of student progress in meeting the learning objectives is part of your duty as the Instructor 1. Classes that last longer than a couple weeks will require that you actively check in with students who are making no progress toward the learning objectives. They may have a plan to complete all of the assignments by the due date. They may not. Check in with them no later than one quarter through the scheduled time frame for the class. Provide feedback as quickly as possible when students submit work.

> *Point of Performance: Understand and meet your defined responsibilities to grade and return student work in a timely manner.*

Let's use a self-paced, 10-week course as an example. Our sample course provides both structure and graduated learning. There are 8 reading assignments, 8 web-based quizzes, and a total of 17 papers due by the course end date. Each section of the course begins with assigned readings defined in 8 blocks.

The readings provide the course content and are the subject of assigned papers. Students read each of the 8 assignments, complete an online quiz, then write a short paper (less than 2 pages) and medium paper (3–4 pages) about each subject. Each paper has reading assignments, learning objectives, reference material guidelines, and length requirements. However, there are no intermediate due dates.

The first 8 weeks of the 10-week course includes 16 assignments broken down as follows: 8 reading assignments, 8 quizzes, 8 short papers, and 8 medium papers. The work is due by the end of week 8. The last assignment is a "final" long paper (8–10 pages) submitted by the end of week 10. Students submit papers and complete quizzes through the AHJ's website with username and password verification. By design, the student should be completing work regularly over the 10-week period assigned for the course. The only time requirement within the AHJ's policy is for the student to complete the first 8 weeks of work, 16 short and medium papers and 8 quizzes, by the end of week 8, allowing 2 weeks for the student to complete the final long paper required in the syllabus. Each short and medium paper, as well as the final long paper, has clearly defined content and quality requirements included in the course syllabus. Each student is required to complete the quizzes and write and submit the assignments in the order outlined in the syllabus.

Since the course work is completely self-paced, and the AHJ has established only two date markers at weeks 8 and 10, the instructor will see that some students use various strategies. Some, likely most, of the students will complete work within the week that it is outlined in the syllabus. Some will complete most, if not all, of the work early in the session. Some students will wait until the deadline to submit. As the Instructor 1, you can only require what is allowed within the policy established by the AHJ. You can, and should, be checking in with students as they complete work to guide and mentor them to improved performance and learning outcomes.

> *Point of Performance: Have a solid understanding of the schedules and expectations outlined by the AHJ. Be ready to meet your responsibilities to grade and return papers within the time frames outlined in the course syllabus.*

The Instructor 1 should communicate regularly with students as they complete their work (communication tools are described later in this chapter). Regularly contact and interact with students as a coach or mentor throughout the process. Though the course is self-paced, it is more likely a student will succeed if submitting work regularly.

> *Point of Performance: Contact any student who has not submitted some work by the end of week two.*

Institution-Paced Courses

Let's discuss institution-paced courses using the same 10-week and 17-assignment course requirements described above except that 8 blog posts (500 words) will take the place of the 8 short papers (2–3 pages).

The institution-based process will require work to be completed each week of the 10-week course. For each of the first 8 weeks, each student will complete the reading, the quiz, and submit a blog post and a medium length paper (3–4 pages). Blog submissions must be posted by midnight on Wednesday each week, and each student must comment on at least three other students' blog posts within the course video conference.

The 8 medium papers are due at the end of each week, commonly at midnight on Sunday. Students must also complete a quiz each week due by midnight on Sunday. Weeks 9 and 10 are set aside to provide students time to write their final long essay (8–10 pages). These weeks generally do not have other required assignments, though students may need to submit an outline of their final paper sometime in week 9. The lack of assignments in the final two weeks allows students to focus their attention on the longer paper.

Grading criteria are similar for both self-paced and institution-paced courses. Students submitting work on time receive full credit. Students submitting late are penalized according to the rules of the AHJ. Commonly a minor (1 or 2 days) delay will have a minor penalty, and a major (3 or more days) delay will have a major penalty. The AHJ will define when work receives no credit even if submitted. Common to both self-paced and institution-paced courses is that the work is generally not accepted at all after the last day.

As an Instructor 1, you must be thoroughly familiar with these requirements and your role within the AHJ to allow discretion. The course syllabus will define the expectations, schedule, grading criteria, and penalties. Instructors are expected to grade papers and maintain alignment with the expectations laid out in the syllabus. Know and follow the rules of the AHJ.

> *Point of Performance: Where you are given the power to use discretion, use it wisely to support students who will benefit from the assistance.*

The Flipped Classroom

Another way to handle online learning is the flipped classroom. The flipped process requires students to access learning content including videos, readings, podcasts, and other self-paced curriculum prior to attending a classroom or training ground session. Classroom sessions may be in person or online but generally includes all the students and the instructor. When the flipped classroom is used in support of a training ground session, all students will likely be required to attend the training ground session.

Classroom sessions focus on clarifying the preclass studies through application-based or discussion-based learning. The process provides students with the opportunity to build and demonstrate mastery of the self-paced material by applying their knowledge in the classroom session. Using the flipped methodology has been shown effective for adult learners, including many individuals who have learning disabilities. Coursework is completed on an institution-based schedule with regular, generally weekly, classroom meeting times.

Learning objectives include the knowledge, skills, and abilities the student will gain through the course. Training ground sessions that are supported by presession flipped classroom study benefit the AHJ by decreasing the amount of time that students must be physically present at the training site. The flipped classroom allows training ground time to be maximized to the physical skills and abilities outlined in the learning objectives. Individuals and companies complete mastery of the knowledge requirement prior to arriving at the training ground. The lesson plan for a flipped classroom should list the training objectives that must be completed prior to the class as Prerequisite Knowledge or Prerequisite Skills.

Computer-Based Training

Computer-based training includes courses where all the learning material is accessed using the computer. The coursework is completed and graded within the system without an instructor having to interact directly with students. Computer-based training is usually self-paced, allowing students to access and complete assignments at their own pace and schedule. AHJs often use the computer-only model to complete legally mandated training. It can also be helpful to ensure students complete the prelearning necessary for a flipped classroom approach. Computer-based training is managed with online learning management systems (LMS), often combined with email notifications and scheduling. The Instructor 1 is not likely to have a role in the computer-based model of learning described here. Job performance requirements (JPRs) of the Instructor 2 and Instructor 3 outline responsibilities for the development, deployment, scheduling, and budgeting for courses destined for a computer-based approach.

The Instructor's Role

When teaching in an online or blended learning environment, you have a variety of tools available to provide coaching, mentoring, and direct instruction to your students. Let's describe some of those tools here, including discussion boards, office hours, video chats, and texting.

Discussion Boards

The discussion board is often used in the electronic learning environment to encourage interaction among the student body and with the instructor. A commonly used format has a different discussion board set aside for each week of the course. This allows the discussion to stay focused on the topic assigned. The LMS provides links for each chapter or section of the designated discussion board. At the top of each discussion board will be the requirements for participating, including an outline of the topic, required length of submission, and reference requirements. Students access the discussion board with their username and password. Each student then writes a blog post according to the outlined expectations and the learning objectives. Once posted, other students and the instructor read and reply with comments on the post. The instructor and other students reply to the submission with supporting or challenging statements based on the curriculum. When students use references in their responses, they must properly cite their references.

Students are generally graded both on the number and quality of posts as outlined in the course syllabus. These requirements encourage interaction and promote high quality, original posts and student responses, creating a positive learning environment online. As the instructor, you will be an active participant in the discussion.

> *Point of Performance: Monitor the tone and tenor of the discussion to ensure it remains a positive learning environment.*

Office Hours

The instructor in an online learning environment must be available to guide, coach, and mentor students through the learning process. Since the participants are in different locations, possibly even different time zones, the AHJ often requires the instructor to identify specific office hours. The office hours give students a defined time that they know the instructor will be available. Communicating during these office hours may take place through a variety of mediums. Online chat, telephone, or video conferencing may all be used depending on the capabilities of the AHJ hosting the class, the student's needs, and the student's access to technology. You may also have the option of directly coordinating office hours at the request of students. Remember, the main goal is to help students achieve mastery of the course objectives.

> *Point of Performance: Effectively communicate availability and the need to coordinate attendance at office hours.*

Video Conferencing

A video chat generally includes each participant having a camera and a computer screen. Video chat sessions can be used effectively for one-on-one, small group, or large group classes or meetings. Applications such as Zoom, Facetime, and Skype are just some of the options available for video chat. The AHJ will likely define the video chat methodology used by an Instructor 1 and will often provide access to the technology. Often nothing more than a smartphone and the corresponding application are required. The variability of the chat environment provides both benefit and challenge for the Instructor 1.

One-on-one sessions using video chat are a modern version of a phone call with the advantage of being able to see the other person. You can often use lesson plan material such as PowerPoint or video during the discussion. These sessions provide a comfortable environment for the individual student. They may be especially effective for students who have challenges communicating or participating in group settings. Video chats also provide an excellent opportunity for individualized instruction. The downside is that they can consume a lot of instructor time.

Small group sessions are effective to spark interaction and discussion in a more "live" environment than that provided in a discussion board format. Small group sessions can be run with or without an instructor present. These sessions provide students with the opportunity to discuss course material in an open learning environment. During these sessions, the person speaking will generally have their icon or video brought forward within the chat application so others can see who is speaking. This also provides a challenge in that individuals may interrupt or attempt to dominate a session.

If the session is instructor-led, then you are responsible to manage the session using techniques like those used in a normal classroom session. One advantage to this format is the "side chat," or direct messaging capability. The side chat allows you to connect directly with an individual student without the others knowing. This allows you to send discreet messages to correct behavior. If a student is dominating or interrupting, you should immediately address the behavior through the side chat portal and follow up with a personal email later. The side chat also allows positive coaching and mentoring. If a student is not participating, use the side chat to prompt entrance into the discussion.

> *Point of Performance: Maintain balance, allowing everyone to participate, and prompt students who are not participating to engage in the discussion.*

Large group sessions on a video chat application can be very similar to a classroom. This type of large group session usually relies on the lecture format as the primary teaching method. Students should be able to see both the instructor, generally in a small window on the side, and the appropriate lesson plan material on the larger screen. Students can submit questions by messaging in a chat box for the instructor. The instructor can plan times to read and address questions (fig. 6–1).

One challenge of the large group video chat is that it can feel impersonal from the student's perspective. This is exacerbated when questions or comments in the chat window are not addressed in a timely fashion.

> *Point of Performance: Answer student questions as soon as practical so students know they are being heard.*

Texting

The use of text messaging is a quick way for students to get questions answered if this is allowed by the AHJ. The use of personal cell phones and exchanging of numbers can be a sticking point for the use of texting. If the AHJ allows texting between students and instructors, it is critical for the instructor to ensure that all messages are course-related and professional.

Fig. 6–1. Lectures via video chat have become increasingly common.

Always understand and operate within the policies of the AHJ hosting a class. Though there can be challenges with the use of texting, many younger students rely on texting as a primary form of communication.

> *Point of Performance: Use whatever tools are available to help students achieve mastery in the course material. If this includes texting, within the rules of the AHJ, use it help your students learn.*

Summary

Computer-based and distance learning are, and will continue to be, a growing part of training in the fire service, particularly since the COVID-19 pandemic. This chapter outlines many of the uses and applications for computer-based and distance learning, including various types of online learning, blended learning, and flipped classrooms. The Instructor 1 must remember that the primary goal of any course is to help students achieve mastery of the learning objectives. These tools can be used individually and in combination to achieve the learning objectives based on the systems and lesson plans designed by the AHJ. As the use of technology expands to encompass more of everyday life, it will also expand both the opportunity and challenge for students and instructors. The focus for the Instructor 1 is to know the roles and responsibilities of the AHJ, thoroughly understand the lesson plan material, have mastery of the technology for each course you teach, and provide the students with a safe and effective learning environment.

EVALUATION AND TESTING

Testing and evaluation are duties for the Fire Service Instructor 1. How you conduct testing varies widely among authorities having jurisdiction (AHJs). Some require registration for exams prior to the first day of class. Others require the instructor to pick up a testing packet the day the test is being administered. There may be a series of procedural requirements from the AHJ, department, county, and state levels. Additionally, passing the test may mean some of your students get reimbursed for a class or even additional pay. All of these are possible needs and outcomes of the testing processes you must be able to perform as an Instructor 1.

Why does the AHJ need testing? Remember that each class you teach will have learning objectives. You are working to ensure that your students achieve the learning objectives as outlined in the lesson plan. Testing and evaluation are the methods used to ensure students can demonstrate mastery of the learning objectives.

It is a duty of the Instructor 1 to conduct testing and evaluation processes as required by the AHJ hosting the class. Additional duties of the Instructor 1 are to ensure that the test is administered in a fair and unbiased manner, that everyone has the same opportunity for success, and that the students perform commensurate with their ability to master the course objectives. The Instructor 1 must be familiar with at least three testing formats: oral, written, and performance-based tests.

The Instructor 1 must also understand the potential for bias in testing, how to grade tests, and how to prepare and submit related paperwork. Let's examine each of these duties and responsibilities together.

Testing Policies and Procedures

Each AHJ hosting a class should have written policies and procedures describing the expectations for instructors and students related to the testing process. For each class you teach, the testing requirements and procedures should be clearly spelled out in the lesson plan provided by the AHJ. Some of these may include checking the student's identification, verifying registration information for the student or department, and accepting or verifying payment for testing services. In addition, you must follow procedures for recording and reporting scores to the AHJ and the student's fire department.

It is critical that you understand testing and evaluation expectations before the class begins.

Point of Performance: Collect and verify testing material, procedures, and expectations well in advance of the testing period. This likely means arriving at the testing site before the testing period is scheduled to begin.

Testing Bias

Testing bias means that students' scores are impacted by things other than their ability to demonstrate mastery of the course material. The differential impact may be based on learner characteristics such as race, gender, or other differentiating group criteria. Testing bias can occur across all three types of testing. This section describes how bias may occur in testing. Recognize the potential for bias, and work diligently to identify and eliminate bias in all testing processes.

Types of Testing

The three testing processes outlined for the Instructor 1 are oral, written, and performance tests. In the following sections, each testing process will be described, including the possible influence of bias in each.

Oral Tests

Oral testing is generally done with one student taking the test at a time. The instructor asks questions, and the student provides answers, demonstrating mastery of the learning objectives. The AHJ defines the learning objectives and the standard for testing. The student is measured based on ability to demonstrate a mastery of the learning objectives, measured against the standard, to identify their score or grade.

One example of the oral test often used in the fire service is the daily check of the self-contained breathing apparatus (SCBA). This test generally evaluates the student's knowledge of the SCBA in combination with their ability to perform the physical skills of the daily check. It is also common for the daily check evaluation to be immediately followed by a timed SCBA donning performance test (fig. 7–1).

The SCBA testing process begins with the student wearing appropriate personal protective equipment (PPE) and having an SCBA. The instructor will have the appropriate grading criteria and a score sheet. The test begins with the instructor asking the student to describe the daily check procedure. The student describes the daily check process and the rationale for each procedure. As a student proceeds, they will generally describe each component of the SCBA in the order outlined on the given daily check form. The instructor listens carefully, taking note of the items the student covers or misses during the daily check. The instructor scores the student based on the student's performance relative to the written standard provided by the AHJ. If allowed by the AHJ, students should be given immediate feedback on the outcome of their test.

Point of Performance: Complete the paperwork for each student immediately.

For an oral test on the SCBA, success is based on the student's ability to remember and describe every critical piece of the SCBA assembly. Some agencies require 100% knowledge of the SCBA as a minimum score due to the critical impact the SCBA has on firefighter safety. For other types of oral testing, a 70% passing score may be acceptable.

Prior to beginning the test, the instructor should ensure the student understands the purpose and scope of the test, the grading criteria, and the measure of acceptable performance. This doesn't necessarily mean showing the score sheet to the student, although, in many AHJs, students have access to the standards for testing. Rather, the procedures defined by

Physical Check	Functional Check
Air Cylinder • Remove cylinder. • Note no level 2 or 3 damage to cylinder. • Ensure cylinder valve outlet, hand wheel, and relief valve are clean and undamaged • Verify hydrostatic test date is within last 5 years. • Ensure pressure is at or above 5,000 psig. • Mount cylinder. Ensure both Snap-Change locks are fully seated (NO ORANGE SHOWING). **Backpack Frame and Straps** • Check for cracks or deterioration in the frame. • Straps move freely, are fully extended, untwisted, and buckles work. • Drop bag attached to left side of frame **Left Side of Air-Pak** • Remove cap from Transfill connection & inspect for signs of damage. Replace cap. • Inspect low pressure hose along left side of bottle & shoulder strap to regulator for damage. • Regulator must be free of moisture and debris. • Ensure regulator gasket is in place around the outlet port. • Verify Purge Knob rotates properly ½ turn & is closed finger tight. • Regulator is stored in belt mounted holder or connected to facepiece. **Right Side of Air-Pak** • Inspect line along right side of bottle & shoulder strap to remote pressure gauge for damage. **Buddy Breathing Hose** • Remove Buddy Breathing hose & cap. Check for damage and replace hose & cap. **Facepiece** • Lens is clean and free of cracks, crazing, melting or other damage. • Ensure all screws are present and tight. • Adjustable head straps work properly, have not deteriorated and are fully extended. • Sizing dots on strap connections are intact. • Inspect Voice emitter ducts for damage. • The nose cup is clean & fully seated between the flanges of the Voice emitter ducts. • Nose cup inhalation valves installed properly.	**Charge the System** • Depress & release air saver switch on regulator. • Open Cylinder Valve fully to charge system. • Listen for Vibralert & Pak-alert alarms. • Hold regulator and ensure all 5 HUD lights illuminate for 20 seconds. • Listen for air leaks around Transfill connection & Buddy Breathing hose. • Ensure that remote pressure gauge reads within 10% psig of the cylinder. **PASS Device** • Leave Air-Pak motionless until warning tones sound and then move the Air-Pak to stop the tones. • Let the PASS go into pre-alarm (20 sec) then full alarm. (+12 sec) and press the YELLOW button twice to reset. • Manually activate the PASS by depressing RED button. • Press YELLOW Button twice to reset. **Facepiece and Regulator** • Don the facepiece. • Check seal by holding palm of hand over inlet connection. Inhale and hold breath for 5 seconds. • Connect regulator and inhale sharply to start air flow. • Operate Purge Knob. • Remove regulator & press air saver switch to reset inhalation valve. • Remove facepiece. **Low Air Alarms** • Close cylinder and check that the alarm pressure gauge does not show a reduction in pressure for 30 seconds due to leaks. • Crack Purge Knob to bleed off air. Verify that HUD lights operate in descending order. • Ensure HUD low air lights flash and the Vibralert sounds at 1/3 cylinder pressure. • Close Purge Knob finger tight. **Place in Service** • Press YELLOW button twice to shut off PASS device. • Store regulator in belt mounted holder or connected to facepiece. • Place Air-Pak and facepiece on apparatus ready to respond.

Fig. 7-1. SCBA check form (courtesy of Seattle Fire Department)

the AHJ should give the instructor confidence that the student has received instruction on the subject being tested and that the testing documents are consistent with the training provided.

> *Point of Performance: Ensure you have mastery of the testing subject and material before you begin. When possible, have another instructor do a mock test with you before the first student candidate arrives at your test site.*

Bias in oral testing may occur as a result of diversity, such as when there are language or cultural differences between the student and the instructor. For example, the student may come from a family background or culture that values speaking quietly. The instructor could interpret this as the student being unsure of the answers. If there are other noises around the testing area, the instructor may not be able to hear the student clearly. The same type of misunderstanding may occur if the student and instructor were born in different countries or even have different accents from different areas of the same country. As the Instructor 1 who is administering the test, it is critical that you take every opportunity to ensure you can hear, understand, and give full credit to the answers the student is providing. Communicate clearly, and ask the student to repeat if you cannot understand. Ensure the test is fair and the student is given full opportunity to get the grade that best represents their mastery of the course objectives.

Written Tests

Written tests in the fire service are generally delivered in the multiple-choice format. Answers are generally recorded on a bubble sheet.

Each question has a number of potential answers listed, most commonly four. The student must decide which answer is most correct and then indicate their choice on the bubble sheet. Since multiple-choice testing is a widely used testing process, most students will be familiar with it. Multiple choice tests are very common for computer-based or distance learning evaluations.

The procedures for written multiple-choice tests generally require the student to fill out test tracking information in a very exact manner. Tracking information may include a course number, date, and test booklet number, plus the name and address of the student.

> *Point of Performance: Carefully walk the entire class through the process of filling out the tracking information before the test begins.*

Another form of written test provides the student with a series of questions the student must answer in paragraph form. The questions for this type of test are generally open-ended and use words such as "describe," "explain," or "summarize information" based on the learning objectives. This type of test can be administered in person or as part of a computer-based or distance learning curriculum. If used in a distance learning format, the testing process often requires students to submit their work through an electronic portal that scans the answers for copied or plagiarized material. A clear standard against which the student's answers can be measured must be provided by the AHJ. The instructor must be familiar with and confident in both the testing process and the course material to ensure proper grading of the student's work.

Written tests can produce bias. Written test bias is often identified after the fact rather than by the Instructor 1 directly. When possible, you should review the written test in advance. Questions and answers should be clearly identifiable from the lesson plan and learning objectives. Actively review the questions and suggested answers for correlation to the instructional material and bias. If you suspect there is bias in the material, address it as soon as possible with the AHJ.

Patterns of bias may appear when tests are examined in a group or over a long period of time. Statistical analysis may indicate a higher failure rate for a specific race, culture, or gender. The AHJ has a duty to ensure that the testing material is fair, objective, and without bias. If you suspect bias may be occurring, you have a duty to report it to the AHJ.

Performance Tests

Firefighting is an endeavor requiring both knowledge and skill. While knowledge can be assessed through written and oral testing, skill must be evaluated through performance tests. Performance tests are used for a wide variety of fire-based skills including placing and climbing ladders, extending fire attack hose, making connections to hydrants, and forcible entry. The performance testing process generally includes the use of a testing skill sheet such as the sample of the testing sheet for a candidate to don the SCBA within one minute found in figure 7–2.

Most fire service skill sheets include a list of equipment that must be present for the student to use during the test. This is followed by a description of the process and the outcome the student must achieve. The sheet will often include a list of step-by-step procedures and relevant safety items, for example: "Given this tool, the student must be able to demonstrate these skills."

TESTING SHEET 10-3	NFPA 1001, 5.3.1	
OBJECTIVE:	Don SCBA in 1 Minute	FEH Chapter: 10
CANDIDATE NAME/NUMBER:		No.:
TEST DATE/TIME		
EQUIPMENT REQUIRED: [Add local requirements if needed]	• SCBA • Structural Firefighting Personal Protective Ensemble	
EVALUATOR INSTRUCTIONS		
CANDIDATE INSTRUCTIONS: NOTE: The evaluator will read the following exactly as it is written to the candidate	The candidate must don and activate the SCBA within 1 minute. The candidate may use the "over-the-head" method or the "coat" method.	
CRITERIA:	NOTE: Based on material from the Skill Drill Instructor Guides	

Critical?			Pass	Fail
		Candidate kneels by the SCBA, places equipment properly.		
		Candidate opens the cylinder valve fully, ensuring that the pressure gauge reads above 90% and that audible and visible alarms activate on the SCBA.		
		The candidate places the SCBA unit on his/her back using either the "over-the-head" or "coat" method.		
		Candidate adjusts the waist strap and shoulder straps for comfort and fit.		
X		Candidate dons the facepiece, tightens the headstraps, checks the seal, and checks the exhalation valve.		
		Candidate inserts the regulator into the mask, inhales to initiate flow.		
		Candidate dons helmet and gloves.		
X		Candidate must have no skin showing.		
		Doff PPE, inspect and place back into service. (This step is not timed.)		

EVALUATOR COMMENTS: [ANY COMMENTS PRO OR CON REGARDING WHAT THE STUDENT ACCOMPLISHED]	
EVALUATOR SIGNATURE:	

Fig. 7–2. Sample skill sheet

The instructor administering the test will ensure the required equipment is present, explain the test and objectives to the student, and observe the student perform the skill. The instructor is looking to ensure the skill is demonstrated effectively, required procedures are followed, and safety is maintained. The instructor documents performance of each procedure and safety item on the skill sheet.

Performance skill sheets are generally pass or fail for each listed skill. Evaluation is then based on the student's performance for each skill against the standard. The skill sheet should clearly indicate which skills are required and which are optional. Performance-based testing generally identifies the procedural steps that must be completed and the specified order for the student to pass. As with other testing, the instructor should ensure that students are familiar with the testing material and grading standard. Good lesson plans combined with good training and motivated adult learners usually result in passing scores for most candidates. Some will fail; most will succeed.

> *Point of Performance: If the first several candidates fail a skill evaluation for the same reason, you should speak with the test control officer or AHJ representative before continuing.*

An example of a performance-based test is the SCBA donning and doffing procedure, often tested following the daily check described above. The instructor would ensure that the student had the required equipment. In this case, that means full structural firefighting PPE, a SCBA, and a facepiece (see fig. 7–2).

The national standard for donning the SCBA is one minute (National Fire Protection Association, 2019). On the signal to begin, the student would don the SCBA completely. Time generally begins when the student touches the SCBA. Time stops when the student indicates they are finished donning the SCBA, commonly by clapping their hands. The instructor watches the student perform the donning procedure while evaluating and noting on the grading sheet the student's performance against the standard. Once the student has completed the donning procedure, the instructor generally completes a "buddy check" to see if the student donned the SCBA properly.

The instructor would then record the performance on the skill sheet and inform the student of the outcome.

> *Point of Performance: Complete the paperwork for the current student and reset the testing station before the arrival of the next student.*

Bias in performance testing can occur in both the design and the administration of the test. The AHJ must ensure that the testing process and the skill sheet procedures are consistent with the performance objectives and standards expected within the AHJ. Bias may occur when the testing and performance standards are set much higher than those required to do the actual work of firefighting. Making physical testing standards higher than those required of the job has been found to produce bias in testing, and agencies have been ordered to change.

Introduction to Evaluation for New Emergency Services Instructors

by Kevin Milan, PhD, CFO[1]

Waiting for the response to find out if a firefighter knows a concept or can perform a skill is inappropriate in training emergency services personnel. The stakes are simply too high. Crews must have confidence that each member knows their job and is able to perform assigned tasks. Simply stated, we must evaluate the competence of our personnel to ensure the concepts covered in training are mastered.

Evaluation and testing are words that traditionally strike fear in the hearts of emergency services students. Performance on a test can mean the difference between being hired, promoted, or passed over. Test anxiety runs high in our students. The "training division" can quickly become the "testing division" if trainers don't grasp the importance of evaluation in the learning process.

How does an instructor, especially a new instructor, battle the anxiety surrounding testing and evaluation? To be successful, make a plan. Starting with the evaluation tool allows the instructor to tailor the learning episode to meet the learning objectives. This strategy can put the students more at ease. What evidence do you need to know that learning occurred and they now possess the knowledge, skills, and abilities (KSAs) you've taught to help them do their job?

It's important that you, as an instructor, know the difference between two very important words in the educational vernacular: evaluation and assessment. These terms are often used interchangeably; however, they are quite different. Knowing the difference between these two terms can improve your instruction and your students' learning. Correct application of these terms can also reduce the anxiety of your students.

"Evaluation" is a broad term reserved for things like programs and systems. We evaluate recruit academies, officer development programs, and training bureaus. "Assessment" is a much more focused term, concentrating on specific KSAs. We assess a firefighter's ability to tie a knot or don a breathing apparatus.

Presenting the concluding activities of a learning episode as an assessment, rather than an evaluation, provides two very important benefits. First, it allows you to partner with learners to increase the probability the learning process is effective. Using the term

(*continued*)

"assessment" suggests to students that you are genuinely interested in ensuring learning is occurring. This concern for appropriate assessment addresses the unique needs of emergency services learners. This can lower the anxiety of your students and increase opportunities for learning.

Assessment instruments in each learning domain are different. Assessing a psychomotor activity or skill is relatively straightforward. When a firefighter correctly ties a knot, the rope handed back to the instructor is hard evidence the skill was performed correctly. Assessing cognitive learning, or book learning, often comes in the form of written instruments. These can include any number of question types. The question type is less important than the review and analysis of student performance on these assessment instruments. A question with low performance by a majority of students may indicate a poorly constructed question, or it may identify that the instructor neglected to cover the material when delivering content.

As tasks become more complicated, combining, in the words of Dr. Harry Carter, the brains and brawn to do the job, students must combine cognitive knowledge and practical skills. These higher level skills require more complex assessments. For example, assessing a recruit throwing a ladder includes a series of tasks and underlying knowledge. The correct ladder must be selected and placed for the assigned task. The recruit must then safely and efficiently place and raise the ladder correctly on level ground, avoiding overhead obstructions.

The raised ladder, with the correct tip placement at the correct climbing angle, is evidence the task was completed. The underlying cognitive elements, or knowledge objectives, may be absent or incorrect even when the final placement is appropriate. Instructors often use skill sheets to identify critical steps required to pass an assessment of tasks that include cognitive and psychomotor objectives. JPRs are the assessment tools most often used to assess psychomotor skills in certification tests.

Knowing the difference between assessment and evaluation is a powerful tool for the new instructor. Disarming the argument that a test is simply a tool to belittle and judge students can dramatically reduce student anxiety. Taking responsibility for your part of the learning contract with your students shows you are invested in them as individuals. This investment shows students you truly care that they are learning the material. We operate in an industry with life and death consequences. It is therefore essential our teaching, and our students' learning, is effective and efficient.

Using the secret weapon of assessment allows an instructor to partner with students to ensure learning is effective. This simple change of terminology can dramatically improve learning. If a responder enters the class anxious and fearing the test, they aren't ready to learn. Stating that learning and teaching will be assessed, rather than evaluated, reduces anxiety and assists the instructor in developing and delivering learning episodes. Whether it's called evaluation, testing, or assessment, this is a critically important step in the learning process.

1. Kevin O Milan serves as assistant chief with South Metro Fire Rescue (CO).

Point of Performance: Before administering a performance-based test, ensure you clearly understand the objectives and what the skill should look like when performed to the standard. Be vigilant in identifying and correcting potential bias as well as reporting it to the AHJ.

Grading and Submitting Paperwork

Grading should always be done in accordance with the standards as outlined by the AHJ. Testing may provide a range of letter grades, may be pass-fail, or may be on a descriptive scale such as poor, good, and great. The Instructor 1 has a duty to understand the grading policies and procedures of the AHJ.

Written tests using bubble sheets will often include a keyed answer sheet for the instructor to grade test papers.

If permitted by the AHJ, let the students review their graded paper and provide them the opportunity to ask about questions they got wrong.

Point of Performance: Use a highlighter or pencil for grading papers.

For oral or performance-based tests, ensure you complete the grading in compliance with the AHJ. Score the test and record the results as indicated in the testing material. If either you or the student feels something is out of order, resolve the problem before the student leaves the testing area. The AHJ should outline and support procedures for the Instructor 1 to follow when there is a disagreement or some other problem in the testing process.

Point of Performance: Know and follow testing procedures as outlined in the lesson plan.

Once the testing is complete, ensure you fill out and submit the testing paperwork in a timely fashion. Grades should be provided as soon as possible. The Instructor 1 also has a duty to ensure that test outcomes are provided to the student privately. The instructor should never discuss a student's outcome with anyone except the student or an AHJ representative.

Point of Performance: Complete and file all required testing paperwork before leaving the testing facility.

Summary

In all testing processes, the Instructor 1 must maintain objectivity and measure the performance against the standard as provided by the AHJ. Testing material must be kept secure until the test is underway. Results should be provided in a timely and confidential manner. Paperwork must be completed accurately, and all records should be submitted immediately after the test is complete.

Reference

National Fire Protection Association. 2019. *NFPA 1001: Standard for Fire Fighter Professional Qualifications.*

INDEX

A

ability
    instructional xv
    learning 40
abuse 37–38
accommodations 31–32
accreditation, minimum 28
ADA (Americans with Disabilities Act) 31
AHJ. *See* authority having jurisdiction (AHJ)
air cylinder 94
air gauge 11
air management training sessions 10
Air-Pak 94
American fire service 54
Americans with Disabilities Act (ADA) 31
apologizing, inappropriate behavior 40
apparatus bay 59–60
application-based learning 83
application, lesson plan 19
area of expertise 6, 38
assessment 97–98
assignments
    lesson plan elements 20

online learning 83–85
papers 81
audiovisual (AV) equipment
    advantages of 64
    disadvantages of 64
    interface tools 59
    job performance requirements (JPRs) and 57–59
auditory learning 24–25
authority having jurisdiction (AHJ)
    assignments and 20
    audiovisual (AV) equipment and 59, 64
    cultural differences and 54
    direct contact for 2, 40
    grading and 101
    inappropriate behavior and 37–38
    learning disabilities and 32
    lectures and 44
    lesson plan and xvii, 1, 5, 73
    online learning and 82–89
    testing and 89–90
    visual aids and 59
autism spectrum disorder 30
AV equipment. *See* audiovisual (AV) equipment

B

behavioral objectives
definition of 5
learning styles and 27
lesson plan and 4–7
student motivation and 8
behavior, disruptive. *See* disruptive
behavior
bias 89–90
in oral testing 95
in performance testing 98–99
in written testing 96
blended learning
styles 24–26, 25–27
blog posts 82, 87
body language 53
bubble sheets 93, 101
Buddy Breathing hose 94

C

cell phones 72
chevron configuration 62
classroom
combined 63
culture 49
environments xvii, 30, 43–44
flipped 83
layout 57
lesson plans 4–5
lighting 70
classroom positioning 57–60
audiovisual (AV) equipment
and 59–60
three-position method and 58
visual aids and 59
classroom setups
Instructor 1 60–61
online classroom 63
traditional classroom 61
U configuration 62–64

V configuration 62–63
coaching 26
definition of 76–78
in online learning 88
sets and reps 13–14
cognitive learning 98
combined classrooms 63
combined learning
styles 25–27, 28
communication 51–53, 82
computer-based training 86–87.
See also online learning
confidence 70
content-based lectures 27
course objectives 23
COVID-19 79
credibility 10, 67
cultural differences
impact in lesson plans 48–54
in learning 29–30
cultural norms
national cultures and 54
student performance and 30
culture 48–54
classroom 49
department 48
of extinguishment 48
of learning 48, 49
national 54–55
of safety 48
of search 48
curriculum
Firefighter 1 23
success 38

D

daily check process 93–95
decoding messages 45
dedication 17
delivery 11

of lesson plan 68
demonstrate one-do one 10–11
demonstrations 11–12, 16, 19
department culture 48
deploying hose 12–13
disabilities 30, 86
discrimination 37–38
discussion-based learning 27, 83
discussion boards 87
disruptive behavior 33
    correction of 36
    side conversations 33
    side talking 33, 37
distance learning. *See* online
        learning
distractions 72, 76
diversity 95
Dixon, Jay 50, 53
documentation xviii
Dodson, Dave 34
drill-ground activity 34
duties xiv–xvi
    online learning 83
    testing 89
dyscalculia 30
dysgraphia 30
dyslexia 30

E
electronic devices 72
embedded videos 74
empathy 51
encoding messages 44
environments
    classroom xvii, 30, 43–44
    hands-on 17
    learning. *See* learning
        environments
    live-fire 71
    simulated-fire 71

zero-visibility 71
evaluation 28
    assessment vs 97–98
    duty xiv
    instruments xvi
    as lesson plan element 20
expectations 48
experience 8
expertise 5, 38
eye contact
    classroom lighting and 71
    classroom positioning and 58
    classroom setups and 63
    cultural differences and 30
    with visual aids 60

F
facepiece 94, 98
Facetime video conferencing 88
fairness 93
FDIC (Fire Department
        Instructors Conference) xix
feedback 19
    four-step communications
        process 46–48
    online learning 81
    teaching 69–70
field of vision 58
Fields, Aaron 38–39
Fire Department Instructors
        Conference (FDIC) xix
*Fire Engineering's Handbook for
    Firefighter I and II* 2
Firefighter 1 23
fire service xx, 38
Fire Service Instructor 1.
    See Instructor 1
first day teaching 69
flat-head axe 11, 14–15
flipped classrooms 83

forcible entry 11, 14
four-step communications
      process 43–45
   decoding 45
   definition 44
   encoding 44
   receiving 45
   transmitting 44
four-step instructional
      method 3, 8
   application 19
   evaluation 20–21
   preparation/motivation 8
   presentation 18
   safety elements 34

G

gender diversity 54
generational differences 55–56
goals 80
grading 85, 98–99
graduated learning 4
group participation 30

H

Halligan tool 11–15
hands-on instructors 9–16, 38–39
   demonstrate one-do one 10–11
   demonstration and 11–12
   environments 17
   having fun 17
   lesson plans 9–10
   positioning 13–14
   positive approach and 15
   prerequisite skills for 15–16
   sets and reps 12
   time management and 16–17
   training with 34
harassment 37–38
having fun 17
hazards 35–36

hearing impairment 30
hose
   deploying 12–13
   loading 4
   picking up 4
   stretching 4
hosting agency. *See* authority
      having jurisdiction (AHJ)
HUD lights 94

I

IIC (instructor in charge) 35
inappropriate behavior 37–38
inclusion 30
individualized instruction 76, 88
information recall 24–25
institution-paced courses 82
instruction
   delivery of xiv–xv
   development of xiv
   individualized 76, 88
   level of 5
   materials 6–7, 19, 84–87
   techniques 18–19
Instructor 1
   certification xiv–xv
   classroom setups 60
   definition of xv–xvi
   purpose of 18
   safety element roles and 35
   student participation and 30
Instructor 2 xiii
   computer-based
      training 86–87
   lesson plan
      development 5, 8, 64
Instructor 3 86
instructor development 66–70
   challenging lesson plan
      sections 69
   PowerPoint teaching and 66

practice 68
preparation 66
self-reflection 70
test teach 69
timing and flow 68
instructor in charge (IIC) 35
instructors 9–16
   approachability and 39
   hands-on 9–16, 38–39
   positioning 13–14, 62, 64
   traits of 52
   values of 50
instructor-to-participant (I/P)
   ratio 35
interface tools 59
internet 44, 48, 79. *See also* online
   learning
I/P (instructor-to-participant)
   ratio 35

J

job performance requirements
   (JPRs) xiii
   as assessment tools 98
   audiovisual (AV)
      equipment and 64
   for classroom setup 57
   for computer-based
      training 86–87
   definition of xiv–xv
   for Instructor 1 xv–xvii
   safety elements and 35
job title 3
JPRs. *See* job performance
   requirements (JPRs)

K

kinesthetic learning 25
knowledge 94, 98
   delivery xv, 18
   transfer 50

knowledge, skills, and abilities
   (KSAs) 97

L

ladders 14, 98
large group sessions 89
late assignments 85
leadership approaches 41
learner characteristics 90
learning 27, 76
   accommodations 32
   aids 18, 64
   application-based 83
   assessment of 97–98
   auditory 24–25
   cognitive 98
   contract 98
   cultural differences and 29
   culture 49
   disabilities 30, 86
   discussion-based 27, 83
   engagement 28
   factors influencing 27–39
   generational differences
      and 55–56
   graduated 4
   kinesthetic 25
   online 79–87
   process 24
   repetition and 28
   visual 24
learning environments xvii, 70
   classroom lighting and 70
   distractions 72
   inclusive 30
   noise and 73
   positive 33, 87
   supportive 16
   teaching objectives and 50
learning management system
   (LMS) 86–87

learning objectives 27–29, 46
  accommodations 32
  classroom environment and 43
  testing and 89
learning pyramid 25
learning styles 23
  auditory 24–25
  combined 24–26, 25–27
  cultural differences and 29
  kinesthetic 25
  visual 24
LEARN qualities 50–53
lecture method 43–44
  cultural differences and 48
  feedback 46
  four-step communications
    process and 43
lectures 18, 25–26
  content-based 27
  format of 43
lesson plans xvi
  accessing 1–2, 4
  adaptation of xvii, 73–75
  challenging sections 69
  components of 2–7
  conflicts 67
  delivery 68, 76
  duties 1
  four-step instructional
    method and 8
  hands-on 9
  review 67, 76
  safety elements 34–36
  summary 19
level of instruction 5
liability releases xviii
life safety, incident stabilization,
    and property conservation
    (LIP) 50
listening 51–52

live-fire
  environments 71
  training 36
LMS (learning management
    system) 86–87
loading hose 4

M

manipulative skills 28
materials, instructional 6–7,
    19, 84–87
media transitioning 73–74
mental models 68–69
mentoring 70
  definition of 76–78
  online learning and 86
messages 44–45
Milan, Kevin 97–98
minimum accreditation 28
mission statements 48
motivation 8–9, 27, 67. *See
    also* preparation
multiple-choice testing 93
muscle memory 11

N

national culture 54–55
National Fire Protection
    Association (NFPA) xiii–
    xiv, 36–37
*NFPA 472: Standard for Competence
    of Responders to Hazardous
    Materials/Weapons of Mass
    Destruction Incidents* 5
*NFPA 1001: Standard for Fire
    Fighter Professional
    Qualifications* xiii, 5, 28–29
*NFPA 1002: Standard for Fire
    Apparatus Driver/Operator
    Professional Qualifications* 5

NFPA 1041 JPRs ix–xi
*NFPA 1041: Standard for Fire and
    Emergency Service Instructor
    Professional
    Qualifications* xiii, 1–2, 8
*NFPA 1403: Standard on Live Fire
    Training Evolutions* 36
noise 73
nozzles 4

O

objectives
    behavioral 8
    course 23
    learning 27
office hours 87
online learning 45, 48, 79–87
    classroom environment for 63
    computer-based training
        and 86–87
    flipped classroom and 83
    institution-paced
        courses and 82
    instructional tools 84–87
    self-paced courses and 80–82
    written testing and 93
open communication 51–52
open-ended testing 96
oral testing 93–96, 102
outlines 10

P

pace of instruction 24–25
    audiovisual (AV)
        equipment and 64
    transmitting messages
        and 45–46
Pak-alert 94
paper assignments 81–82
paperwork xviii, 93, 102

participation, group 30
PASS (personal alert safety
    system) device xviii, 94
performance, student 30
performance testing 94, 102
personal alert safety system
    (PASS) device xviii, 94
personal protective equipment
    (PPE) xviii, 15, 34, 93
physical safety 54
picking up hose 4
plagiarism 96
positioning
    classroom 57–60
    instructor 13–14
positive approach 15
positive influence 77
positive learning environments.
    See learning environments
post-session analysis 39
posttraumatic stress disorder
    (PTSD) 30
PowerPoint 45, 64
    media transitioning and 73
    teaching with 66
PPE. *See* personal protective
    equipment (PPE)
practice 68
preparation xv, 8–9, 49. *See
    also* motivation
    feedback and 46
    hands-on training and 39
    instructor development and 66
    lesson plans and 73
prerequisite skills 15
presentation
    of knowledge 18
    of skills 19
pressure gauge 11
pretraining planning process 15
program management xiv

psychological safety 54
psychomotor activities 100–101
PTSD (posttraumatic stress
        disorder) 30
Purge Knob 94

Q
questions 46
    as an assessment
        instrument 98
    about lesson plan 67–69
    from students 30, 89
quizzes 81–82

R
radios 72
rapport 49
reasonable
        accommodations 31–32
receiving messages 45
reference materials 67
references 8
repetitions
    hands-on instruction and 39
    in learning 28
reporting requirements 40
requirements, minimum xvii
respect 51
roof operations 14–15

S
safe incident handling 34
safety xvii
    elements 34–36
    lighting and 71
    physical 54
    psychological 54
    structural firefighter bunking
        gear and 29
safety officer (SO) 35

SCBA. *See* self-contained breathing
        apparatus (SCBA)
score sheet 93
scrimmage 39
Seattle (WA) Fire
        Department xix, 39
self-contained breathing
        apparatus (SCBA) xviii
    checking 11, 94
    donning and doffing
        procedures 98
    oral test 93
self-paced courses 80–82
self-reflection 70
sessions
    large group 89
    small group 88
    training xvi, 38
sets and reps 12
side talking 33
simulated-fire environments 71
skills 9, 94, 98
    building 10–11
    manipulative 28
    prerequisite 15
skill sheets 94–95
Skype 88
small group sessions 88
Snap-Change locks 94
SO (safety officer) 35
standard operating guidelines
        (SOG) 67–68
storytelling 67
stretching hose 4
structural firefighter bunking
        gear 28
students 12–16, 23
    with disabilities 30–32
    disruptive behavior and 33
    engagement of 68
    motivation of 27

on duty 72
  self-paced courses and 84
  traits 23
  types of 73
subject matter expertise 5
syllabus 85

T

teaching xvi, xix
  aids 24
  as an art 50
  combined learning styles 25
  discussion-based 27
  generational differences and 55
  learning styles 23
  methodologies 57
  objectives 4
  persona 53
  skill development and 9, 66
  styles xvii, 25, 63
  systematically 38
  with visual aids 60
test anxiety 97
test control officer 98
testing 20, xiv
  bias 90–96
  duties 89
  multiple-choice 93
  open-ended 96
  oral 93–96, 102
  policies 92–93
  written 93, 101
testing formats 89–90
  oral 93, 102
  performance 94, 102
  written 93, 101
test teach 69
texting 89
three-position method 58, 61
time management 16

time standards 28–29
tone of voice 24–25, 45–46
tools
  flat-head axe 11, 14–15
  Halligan 11–15
  hose 4, 12
  instructional 6–7, 19, 84–87
  nozzles 4
traditional classrooms 61
training
  activities 34–35
  hands-on 10, 38
  safety 34–36
  sessions xvi, 38–39
  time management and 17
training ground
  culture 49
  flipped classroom and 83
  learning environment xvii
  lesson plans 4–5
  noise and 73
  repetition in learning and 28
transmitting messages 44
traumatic brain injury 30
troubleshooting 65
trust 10, 51, 76

U

U configuration 62–64
United States Department of
    Labor 37
U.S. Fire Administration 36

V

value 68
V configuration 62–63
vertical ventilation 14
Vibralert 94
video conferencing 88–89
videos 74

virtual classroom. *See* online
　　learning
Viscuso, Frank 77–78
visual aids 24, 52, 53
　　classroom positioning
　　　　and 59–60
　　classroom setup and 62
　　receiving messages and 45
visual impairment 30
visual learning 24

volume of voice 45–46, 60

W

web-based instruction. *See* online
　　learning
written testing 93, 101

Z

zero-visibility environments 71
Zoom video conferencing 88